SpringerBriefs in Computer Science

SpringerBriefs present concise summaries of cutting-edge research and practical applications across a wide spectrum of fields. Featuring compact volumes of 50 to 125 pages, the series covers a range of content from professional to academic.

Typical topics might include:

- A timely report of state-of-the art analytical techniques
- A bridge between new research results, as published in journal articles, and a contextual literature review
- A snapshot of a hot or emerging topic
- An in-depth case study or clinical example
- A presentation of core concepts that students must understand in order to make independent contributions

Briefs allow authors to present their ideas and readers to absorb them with minimal time investment. Briefs will be published as part of Springer's eBook collection, with millions of users worldwide. In addition, Briefs will be available for individual print and electronic purchase. Briefs are characterized by fast, global electronic dissemination, standard publishing contracts, easy-to-use manuscript preparation and formatting guidelines, and expedited production schedules. We aim for publication 8–12 weeks after acceptance. Both solicited and unsolicited manuscripts are considered for publication in this series.

**Indexing: This series is indexed in Scopus, Ei-Compendex, and zbMATH **

More information about this series at https://link.springer.com/bookseries/10028

Youyang Qu · Longxiang Gao · Shui Yu ·
Yong Xiang

Privacy Preservation in IoT: Machine Learning Approaches

A Comprehensive Survey and Use Cases

 Springer

Youyang Qu ⓘ
Data61
Australia Commonwealth Scientific
and Industrial Research Organization
Melbourne, VIC, Australia

Shui Yu ⓘ
School of Computer Science
University of Technology Sydney
Ultimo, NSW, Australia

Longxiang Gao ⓘ
Shandong Computer Science Center
Qilu University of Technology
Shandong, China

Yong Xiang ⓘ
School of Information Technology
Deakin University
Burwood, VIC, Australia

ISSN 2191-5768   ISSN 2191-5776 (electronic)
SpringerBriefs in Computer Science
ISBN 978-981-19-1796-7   ISBN 978-981-19-1797-4 (eBook)
https://doi.org/10.1007/978-981-19-1797-4

This Springer imprint is published by the registered company Springer Nature Singapore Pte Ltd.
The registered company address is: 152 Beach Road, #21-01/04 Gateway East, Singapore 189721, Singapore

Preface

Internet of Things (IoT), as a booming computing architecture, is experiencing rapid development with a speed beyond imagination. Nowadays, IoT devices are so pervasive that they have become key components of human daily life, such as sensors, intelligent cameras, smart wearable devices, and a lot more. By reshaping the existing network architecture, IoT has provided significant convenience and improvement of quality of life.

Since IoT devices are deployed ubiquitously, an increasing volume of data is collected and transmitted over IoTs. The statistic shows total data volume of connected IoT devices worldwide is forecast to reach 79.4 zettabytes (ZBs) by 2025. However, the data privacy issues become even severe because sensitive information of collected data is not properly managed, especially health data, location data, identity-related data, etc. Moreover, data from multiple sources pose further challenges since the interconnections among the data may reveal more sensitive information. Furthermore, the advancement of data pattern extraction and data analysis techniques put privacy under more serious threats. Thus, privacy preservation has become a crucial issue that needs to be well considered in this age of IoT.

Machine learning has proved its superior performance in data manipulation field. In addition to perform predictive analysis or optimization-oriented services, machine learning algorithms are adopted in privacy-preserving data sharing and publishing scenarios. It attracts extensive interest from both academia and industry. Among all existing solutions, reinforcement learning, federated learning, and generative adversarial networks (GAN) are the most popular and practical ones. Extensive research has been conducted to leverage or modify them for privacy protection considering diverse conditions. Therefore, they are also the main focus of this monograph, through which the rationale of machine-learning-driven privacy protection solutions are present.

In this monograph, we are going to comprehensively and systematically introduce machine-learning-driven privacy preservation in Internet of Things (IoTs). In this big data era, an increasingly massive volume of data is generated and transmitted in IoTs, which poses great threats to privacy protection. Motivated by this, an emerging research topic, machine-learning-driven privacy preservation, is fast booming to address various and diverse demands of IoTs. However, there is no existing literature discussion on this topic in a systematical manner. The authors in this monograph aim to sort out the clear logic of the development of machine-learning-driven privacy preservation in IoTs, the advantages, and disadvantages of it, as well as the future directions in this under-explored domain. The issues of existing privacy protection methods (differential privacy, clustering, anonymity, etc.) for IoTs, such as low data utility, high communication overload, and unbalanced trade-off, are identified to the necessity of machine-learning-driven privacy preservation. Besides, the leading and emerging attacks pose further threats to privacy protection in this scenario. To mitigate the negative impact, machine-learning-driven privacy preservation methods for IoTs are discussed in detail on both the advantages and flaws, which is followed by potentially promising research directions.

The prominent and exclusive features of this book are as follows:

- Reviews exhaustive the key recent research into privacy-preserving techniques in IoTs.
- Enriches understanding of emerging machine learning enhanced privacy-preserving techniques in IoTs.
- Covers several real-world applications scenarios.
- Maximize reader insights into how machine learning can further enhance privacy protection in IoTs.

This monograph aspires to keep readers, including scientists and researchers, academic libraries, practitioners and professionals, lecturers and tutors, postgraduates, and undergraduates, updated with the latest algorithms, methodologies, concepts, and analytic methods for establishing future models and applications of machine-learning-driven privacy protection in IoTs. It not only allows the readers to familiarize with the theoretical contents but also enables them to make best use of the theories and develop new algorithms that could be put into practice.

The book contains roughly three main modules. In the first module, the book presents how to achieve decentralized privacy using blockchain-enabled federated learning. In the second module, the personalized privacy protection model using GAN-driven differential privacy is given. In the third module, the book shows the hybrid privacy protection using reinforcement learning. Based on the above knowledge, the book presents the identified open issues and several potentially promising future directions of personalized privacy protection, followed by a summary and outlook on the promising field. In particular, each of the chapter is self-contained for the readers' convenience. Suggestions for improvement will be gratefully received.

Melbourne, Australia   Youyang Qu
Shandong, China   Longxiang Gao
Ultimo, Australia   Shui Yu
Burwood, Australia   Yong Xiang

Acknowledgments

We sincerely appreciate numerous colleagues and postgraduate students at Deakin University, Melbourne and University of Technology Sydney, Sydney, who contribute a lot from various perspectives such that we are inspired to write this monograph. We would like to acknowledge the support from the research grant we received, namely, ARC Discovery Project under the file number of 200101374. In this book, some interesting research results demonstrated are extracted from our research publications that indeed (partially) supported through the above research grants. We are also grateful to the editors of Springer, especially Dr. Nick Zhu, for his continuous professional support and guidance. Finally, we would like to express our thanks to the family of each of us for their persistent and selfless supports. Without their encouragement, the book may regrettably become some fragmented research discussions.

Melbourne, Australia   Youyang Qu
Shandong, China   Longxiang Gao
Sydney, Australia   Shui Yu
Melbourne, Australia   Yong Xiang
December 2021

Contents

1   **Introduction**   1
     1.1   IoT Privacy Research Landscape   1
     1.2   Machine Learning Driven Privacy Preservation Overview   2
     1.3   Contribution of This Book   3
     1.4   Book Overview   4

2   **Current Methods of Privacy Protection in IoTs**   7
     2.1   Briefing of Privacy Preservation Study in IoTs   7
     2.2   Cryptography-Based Methods in IoTs   9
     2.3   Anonymity-Based and Clustering-Based Methods   11
     2.4   Differential Privacy Based Methods   13
     2.5   Machine Learning and AI Methods   14
         2.5.1   Federated Learning   15
         2.5.2   Generative Adversarial Network   16
     References   16

3   **Decentralized Privacy Protection of IoTs Using**
    Blockchain-Enabled Federated Learning   19
     3.1   Overview   19
     3.2   Related Work   21
     3.3   Architecture of Blockchain-Enabled Federated Learning   23
         3.3.1   Federated Learning in FL-Block   23
         3.3.2   Blockchain in FL-Block   24
     3.4   Decentralized Privacy Mechanism Based on FL-Block   27
         3.4.1   Blocks Establishment   27
         3.4.2   Blockchain Protocols Design   29
         3.4.3   Discussion on Decentralized Privacy Protection Using
               Blockchain   29
     3.5   System Analysis   30
         3.5.1   Poisoning Attacks and Defence   30
         3.5.2   Single-Epoch FL-Block Latency Model   31
         3.5.3   Optimal Generation Rate of Blocks   34

3.6   Performance Evaluation 35
      3.6.1   Simulation Environment Description 35
      3.6.2   Global Models and Corresponding Updates 37
      3.6.3   Evaluation on Convergence and Efficiency 38
      3.6.4   Evaluation on Blockchain 41
      3.6.5   Evaluation on Poisoning Attack Resistance 42
3.7   Summary and Future Work 45
References 45

4   Personalized Privacy Protection of IoTs Using GAN-Enhanced
Differential Privacy 49
4.1   Overview 50
4.2   Related Work 51
4.3   Generative Adversarial Nets Driven Personalized Differential
      Privacy 53
      4.3.1   Extended Social Networks Graph Structure 53
      4.3.2   GAN with a Differential Privacy Identifier 54
      4.3.3   Mapping Function 57
      4.3.4   Opimized Trade-Off Between Personalized Privacy
              Protection and Optimized Data Utility 61
4.4   Attack Model and Mechanism Analysis 62
      4.4.1   Collusion Attack 62
      4.4.2   Attack Mechanism Analysis 63
4.5   System Analysis 64
4.6   Evaluation and Performance 65
      4.6.1   Trajectory Generation Performance 67
      4.6.2   Personalized Privacy Protection 67
      4.6.3   Data Utility 69
      4.6.4   Efficiency and Convergence 69
      4.6.5   Further Discussion 71
4.7   Summary and Future Work 73
References 74

5   Hybrid Privacy Protection of IoT Using Reinforcement
Learning 77
5.1   Overview 78
5.2   Related Work 80
5.3   Hybrid Privacy Problem Formulation 80
      5.3.1   Game-Based Markov Decision Process 80
      5.3.2   Problem Formulation 81
5.4   System Modelling 82
      5.4.1   Actions of the Adversary and User 82
      5.4.2   System States and Transitions 83
      5.4.3   Nash Equilibrium Under Game-Based MDP 84

   5.5   System Analysis 86
       5.5.1 Measurement of Overall Data Utility 86
       5.5.2 Measurement of Privacy Loss 86
   5.6   Markov Decision Process and Reinforcement Learning 88
       5.6.1 Quick-Convergent Reinforcement Learning Algorithm 88
       5.6.2 Best Strategy Generation with Limited Power 89
       5.6.3 Best Strategy Generation with Unlimited Power 90
   5.7   Performance Evaluation 91
       5.7.1 Experiments Foundations 92
       5.7.2 Data Utility Evaluations 92
       5.7.3 Privacy Loss Evaluations 96
       5.7.4 Convergence Speed 102
   5.8   Summary and Future Work 104
   References 106

6 Future Research Directions 111
   6.1   Trade-Off Optimization in IoTs 111
   6.2   Privacy Preservation in Digital Twined IoTs 112
   6.3   Personalized Consensus and Incentive Mechanisms
       for Blockchain-Enabled Federated Learning in IoTs 113
   6.4   Privacy-Preserving Federated Learning in IoTs 113
   6.5   Federated Generative Adversarial Network in IoTs 114

7 Summary and Outlook 117

Chapter 1
Introduction

1.1 IoT Privacy Research Landscape

IoT is experiencing fast proliferation with advancement of computation and communication technologies like edge computing and 5G. Along with this, a mass volume of data is collected and transmitted over IoT, and thereby raise privacy concerns. Sensitive information in the collected data are usually improperly treated without privacy-preserving techniques. Furthermore, data from different sources may be linked to reveal more sensitive information. All these issues put privacy protection under great threats.

To mitigate privacy leakage, researchers have investigated into various techniques and application scenarios. Before jumping into the privacy research overview, it would be helpful to distinguish privacy from security, which is also shown in Fig. 1.1. Usually, typical security cases include three parties, which are data sender, data receiver, and adversaries. However, privacy protection normally assumes that only two parties are involved, including data curator and data requester. A data curator will not try to distinguish reliable data requester from malicious data requester. The privacy-preserving data is usually distorted using certain techniques, such as generalization, compression, clustering, etc. The primary target of privacy research is obtaining the optimal trade-off between privacy protection and data utility.

Diverse privacy preservation solutions are proposed in recent years leveraging all kinds of techniques. Typically, privacy preservation solutions include clustering-based methods, differential privacy based methods, and cryptography methods. The most well-known clustering-based methods including K-anonymity, L-diversity, and T-closeness. They consider the size consistency, diversity consistency, and distribution consistency, respectively. Differential privacy is a probabilistic model that ensures privacy protection with strict mathematics derivation. Its extensions and variants are used in a wide range of applicatoin scenarios. Cryptography methods include but not limited to homomorphic encryption, which enables privacy protection by operations on the cipher text instead of plain text.

© The Author(s), under exclusive license to Springer Nature Singapore Pte Ltd. 2022   1
Y. Qu et al., *Privacy Preservation in IoT: Machine Learning Approaches*,
SpringerBriefs in Computer Science, https://doi.org/10.1007/978-981-19-1797-4_1

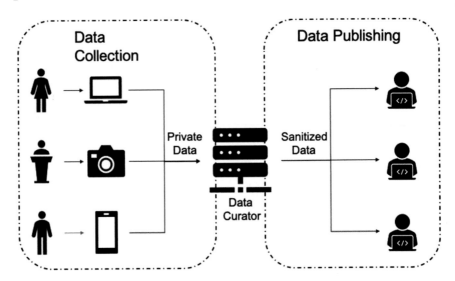

Fig. 1.1 Privacy-preserving data publishing and sharing in IoTs

Due to low scalability, clustering-based methods are not applicable in IoT scenarios. Besides, homomorphic encryption requires how computing power, which is also not feasible in the resource-constrained situations. The adoption of differential privacy in IoT scenarios is well-investigated, but there are certain limitations, for example, low data utility. Therefore, new techniques, especially machine learning based ones, are devised to further optimize the trade-off between privacy protection and data utility.

1.2   Machine Learning Driven Privacy Preservation Overview

The integration of machine learning to privacy preservation solutions is a hot topic in recent years. There are two main categories. The first one is to design privacy-preserving machine learning models, such as generative adversarial network and its variants. The second one is to develop privacy-preserving machine learning paradigms, like federated learning and its extensions.

Generative adversarial network (GAN) is first proposed to generate synthetic images to best mimic the raw image. This brings it a nature of privacy protection. Therefore, it is then used to generate synthetic data to achieve privacy protection with high data utility. The synthetic data generation follows a gaming process. There are two perceptrons, which are Generator and Discriminator. Generator generates data randomly constrained by feedback of Discriminator. Discriminator distinguishes the generated data from raw data and then provides feedback to the Generator. After finite iterations, when the Discriminator cannot distinguish the generated data from raw data, the training of the model is finished. The generated data is regarded as privacy-preserving synthetic data.

As for federated learning, it is a machine learning paradigm that can accommodate almost all machine learning models. The classic machine learning model training requires the collection of data to a centralized server. Consequently, the privacy of data may be breached by the server or during the transmission process. To solve this issue, federated learning trains the models in a distributed way. Usually, a central server will distribute a warm-up model to all local devices. Each of the local devices train the model using locally stored data and then only return the model parameters to the central server. The central server then aggregates all local model parameters and distribute the global model parameters for training of next iteration until convergence.

There are also other machine learning models can be leveraged to preserve privacy. The integration is fast increasing in volume and diversity. This monograph presents the most popular ones that may provide the readings insights on developing new machine learning driven privacy protection solutions, from both perspectives of algorithms and paradigms.

1.3   Contribution of This Book

In this book, we are going to comprehensively and systematically introduce machine learning driven privacy protection in IoT scenarios. With the popularization of IoT, an ever-growing amount of data is collected and processed, which brings about privacy concerns. Traditional methods like differential privacy has its limitations when adopting into IoT privacy protection. Inspired by this, machine learning driven privacy protection solutions are developed to fill this gap. Nevertheless, there is not existing literature discussing machine learning driven privacy protection in IoT in a comprehensive manner. To mitigate the fragmentary knowledge issue in this domain, the authors target on clarity the logic of the development process of machine learning driven privacy protection. We further provide the evaluations on advantages and disadvantages along with potentially promising future directions of this under-investigated field.

To show the outline of this monograph, we follow the order of introduction, current methods of privacy protection in IoTs, decentralized privacy protection of IoTs using blockchain-enabled federated learning, personalized privacy protection of IoTs using GAN-enhanced differential privacy, hybrid privacy proteciton in IoTs using reinforcement learning, and potentially promising future directions. We identify several privacy concerns in existing methods (e.g. cryptography, differential privacy, machine learning, etc.), for instance, data utility degradation and low strictness. These concerns confirms the necessity of machine learning driven privacy protection. To alleviate these concerns, machine learning driven privacy protection is correspondingly articulated from both aspects of benefits and drawbacks. Classic privacy-preserving methods, like anonymity-based methods and differential privacy based methos are analyzed in a comparative and intersectional way versus emerging methods like federated learning and generative adversarial network.

To make this monograph easy to follow, we present most of the theoretical and experimental results in the formats of tables, figures, diagrams, etc. With these presentation skills, we aim to provide the readers the general framework and research status of topic as well as exclusive standpoints on general knowledge and technical insights. Since each of the chapter is self-contained, the readers are allowed to start from anywhere to investigate machine learning driven privacy protection in IoTs. We hope the uniqueness of this monograph will pave a way for the force-coming readers and researchers. We encourage the readers to acquire some basic background knowledge on machine learning and differential privacy. Some concepts and theories are briefly explained to the readers in each chapter, which will help readers with getting a hang of the topic.

This monograph covers a wide range of knowledge, theories, and tools in this topic. Therefore, it is able to serve as a reference for intrigued academics, engineers, or students to gain a rough idea of research status and skills of machine learning driven privacy protection in IoTs. At the same time, this monograph can be used as a textbook for educators and undergraduate who are attracted by this topic.

1.4   Book Overview

The outline of this book with brief introduction is shown as follows.

In Chap. 2, we present current methods of privacy protection in IoTs, including cryptography based methods, differential privacy based methods, anonymity-based methods, and clustering-based methods.

In Chap. 3, we introduce decentralized privacy protection of IoTs using blockchain-enabled federated learning. The integration of federated learning and blockchain provides decentalized protection to sensitive information while the reliability against attacks is largely improved.

In Chap. 4, we show personalized privacy protection of IoTs using GAN-enhanced differential privacy. By adding an extra perceptron, namely, differential privacy identifier, the integration of GAN and differential privacy enables strict personalized privacy protection with improved data utility. The trade-off is correspondingly optimized.

In Chap. 5, we demonstrate the hybrid privacy protection in IoTs using reinforcement learning. The data curator and data requester are modelled in a multi-stage game by properly defining actions, system states, payoff functions, etc. The multi-stage game is then solved by a modified reinforcement learning model, with which the Nash Equilibrium denotes the best privacy protection policies.

In Chap. 6, we provide several potential promising research directions including trade-off optimization in IoTs, privacy preservation of digital twins in IoTs, privacy-preserving blockchain-enabled federated learning in IoTs, privacy-preserving federated learning in IoTs, federated generative adversarial network in IoTs. We aim to provide the readers and forthcoming researchers some insights of this under-investigated field.

In Chap. 7, we summarize and conclude this book.

Chapter 2
Current Methods of Privacy Protection in IoTs

In this chapter, we present the mainstream research of current privacy preservation in IoTs built upon the literature review we have done in recent years [1–3]. Besides, we also summarize several of our book collections to show the frontier of privacy preservation research in edge computing [4–9] and privacy preservation research in edge computing in big data [10–14]. Existing privacy preservation techniques in IoTs include cryptography based methods, differential privacy methods, anonymity-based methods, and clustering-based methods.

2.1 Briefing of Privacy Preservation Study in IoTs

In this section, we show a generalized privacy-preserving system in IoTs, along with roles of each participating party, operations, and status of data. To better clarify, we discuss the definitions, preliminaries, and terms of the generalized system.

From the perspective of participants, there are totally four different roles corresponding to each participating part, which are data generator, data curator, data requester, and data attacker.

- Data generator: Parties (e.g. sensors, individuals or institutions) that generates or collects raw data, such as census data, bank statements, etc. The provide data to other parties in both active and passive ways like uploading posts in social networks or filling in registration forms for specific services.
- Data curator: Parties (e.g. individuals or institutions) that manage the lifecycle of data including collection, storage, and publishing or sharing of the data. At least, the publishing or sharing of data requires anonymization.

© The Author(s), under exclusive license to Springer Nature Singapore Pte Ltd. 2022
Y. Qu et al., *Privacy Preservation in IoT: Machine Learning Approaches*,
SpringerBriefs in Computer Science, https://doi.org/10.1007/978-981-19-1797-4_2

- Data requester: Parties (e.g. individuals or institutions) that are willingly to access the published data for proper uses including business analytics, predictive services, etc.
- Data attacker: Parties (e.g. individuals or institutions) that perform malicious operations to gain more sensitive information from the published data for vicious purposes. It should be noticed that the data attacker is a subset of the data requester.

Usually, we consider three main basic data operations in privacy-preserving IoT systems as follows.

- Data collection: Data generators and data curators collect data from multiple raw data sources.
- Data anonymization: Data curators perform anonymization on the collected raw data so that the ananoymized data can be published or shared to the public.
- Data request: Data requester including data attackers directly access or conduct data retrieval on the published data.

Generally speaking, the lifecycle of a dataset of a IoT system has thee different statuses as below.

- Raw data: The data in its original format before collection
- Collected data: The data processed, transmitted, and stored in the database of data requests before other operations.
- Anonymized data: The data processed by anonymization, generalization, suppression, etc.

A data attacker may launch attacks to any of the operations or roles to further breach the sensitive information. This puts the privacy preservation under great risks. That's also the reason why privacy leakage attracts increasing public concerns.

As to each piece of data, which is also know as record, there are four main types of attributes. Some of them are critical to privacy preservation while some of them has little impact. The details are as follows.

- Explicit identifier: One or several exclusive attributes that uniquely linked to an individual or identity like name, passport ID, social benefit number, etc.
- Quasi-identifier: The attributes that can be used to identify an individual by a combination of several or all of them with a high probability. The most well-known quasi-identifies include birthday, age, postal address, etc.
- Sensitive attributes: The sensitive information that adversaries are interested in, such as salary, medical information, daily routines, etc. In most cases, the values of the sensitive information are unpredictable. That's why data attackers make best efforts to steal them using explicit identifiers or quasi-identifiers.
- Non-sensitive attributes: The attributes that falls out of the previous three categories.

Table 2.1 An instance table of census data

Name	ID	Job	Gender	Age	Salary	Other
Alice	123ABC	Lecturer	F	30	$ 82,000	
NA						
Bob	345DEF	Plumber	M	40	$ 76,000	NA
...	...	...	...	...	...	...

Usually, quasi-identifiers are background knowledge of data attackers. Therefore, in addition to pure background knowledge attacks, data attackers also launch collusion attacks and linkage attacks by collude with other data attackers or link data from multiple sources.

To make it easy to follow, we show an instance of census data in Table 2.1. In this instance, *name* and *ID* are explicit identifiers. Then, we can tell *job*, *gender*, and *age* form a group of quasi-identifiers, which is followed by *salary* as the sensitive attribute. The rest is a non-sensitive attribute, which has little impact on the dataset in terms of privacy preservation.

2.2 Cryptography-Based Methods in IoTs

Cryptography-based methods has been developing for decades for security and privacy protection. Nowadays, cryptography-based methods is still one of the dominant privacy preservation techniques in IoT scenarios.

Cryptography-based methods has been deployed in a wide range of application scenarios. For instance, the medical record of a patient is encrypted with the public key of his/her general practitioner and then the ciphertext is stored into the corresponding database. In this process, the privacy of the patient is strictly preserved against any data relay attackers.

While entering the big data era, IoTs plays an increasingly important role to collect and manage data. For example, the smart wearable devices collects individual health information for analysis and prediction. However, the health data should only be accessed by authorized parties like general practitioners or health consulting agencies. In this case, public key encryption is not feasible because the conflict of large number of requests and limited resources of IoT devices poses significant challenges to key management issues. Motivated by this, Attribute-Based Encryption (ABE) is proposed in 2004 to improve efficiency [15–17]. Under the framework of ABE, a group of descriptive attributes, like identification number of a doctor, can be used to encrypt transmitted data by generating a private key. Then, the ciphertext can be decrypt if the attributes of the user key and ciphertext match. In this way, ABE successfully integrates access control into encryption and subsequently solve the problem of key exchange.

Cryptography-based privacy preservation methods faces a dilemma in IoTs. For one thing, sufficient privacy protection should be provided to users. For another, the ciphertext should be meaningful and informative for data analytics. Consequently, there are several challenges as follows.

The first challenge is information retrieval on ciphertext, namely, encrypted data. The research to mitigate this challenge is known as searchable encryption, which boomed around 2000 [18, 19]. The data is in the form of document in this scenario. The data curator encrypts the indexed document collection, after which the index and encrypted data are stored locally or uploaded to a cloud server that may either be trustful or malicious. When a data user tries to access the data, the data curator uses his/her keyword to generate a trapdoor key, with which the server can recover a pointer to the encrypted documents.

Another representative challenge is direct operation on ciphertext. The research to mitigate this challenge is known as homomorphic encryption proposed in 1978 [19]. Homomorphic encryption is expected to perform pre-designed computing tasks directly on ciphertext and generate an encrypted output. Then, the decrypted output should match the output of direction operations on plaintext. To better clarify, we present it in a mathematical way. Given a piece of data d, a key k, as well as an encryption algorithm $E(\cdot)$, a ciphertext can be described as $E_k(d)$. Moving on, given a function f and its corresponding function is f', $D_k(d)$ as the decryption algorithm under key k, an encryption scheme is regarded as a homomorphic one when $f(m) = D_k(f'(E_k(m)))$ applies.

This type of homomorphic encryption is subject to several pre-designed operations, which limits its application. To improve it, Gentry et al. developed a Fully Homomorphic Encryption (FHE), which supports arbitrary computation on ciphertexts in 2019 [20]. Interested readers are encouraged to explore a survey conducted in [21]. However, a fully homomorphic encryption system is still impractical because of the significant cost in computation power. A variant of FHE entitled Multi-Party Computation (MPC), which was proposed by Yao et al. in 1982 [22], is widely deployed in real-world scenarios. MPC is more efficient while sacrificing a bit security guarantees. The methodology of MPC is kind of distributed computing. Several participants derive a public function with joint efforts based on each participant's local input while the input privacy can be preserved against the others.

It should be noted that encryption can preserve the privacy of an physical object, but is vulnerable against side information attacks, which is also a leading attack in privacy-preserving domain, especially in IoTs. An instance would be network traffic analysis attacks in fully-anonymous communication systems. To be more specific, encryption can protect a web page but is unable to hide the fingerprints of the web page, such as number of web objects, size of the text, sizes of web objects, etc. Based on certain background knowledge, a data attacker is able to find out the visited web pages or websites using network traffic analysis techniques [23–25]. Currently, several solutions have been carried out, among which the most popular one is information theory based packet padding, in particular, dummy packet padding [26] and predicted packet padding [27].

2.3 Anonymity-Based and Clustering-Based Methods

Anonymity-based and clustering-based methods are traditional privacy-preserving methods. These two methods will distort the data in a randomized but controllable way, which should be done by data curators. The data received by the data requesters and data attackers cannot fully restore the data but can mine the data utility from perspectives of statistics or patterns.

The most representative anonymity-based and clustering-based methods are k-anonymity, l-diversity, and t-closeness. Table 2.2 is used to show the development course of data clustering methods for privacy protection in IoTs.

K-anonymity is proposed in 1998, which requires each record in a cluster is at least identical to the other $k - 1$ records on the quasi-identifier. As a probabilistic model, k-anonymity sharply decreases the probability of a record being re-identified to $\frac{1}{k}$. It also performs data generalization. As Table 2.2 shows, like dancers, singers, etc. are denoted by artists. In addition to the textual data, numeral data are expressed in the form of a range, such as the age column. In this case, the value of k is 2 and the re-identification probability is $\frac{1}{2}$. If the cluster size is increased, the value of k is great enough to preserve each individual's privacy. The mathematical description is as follows.

Let $T = t_1, t_2, ..., t_n$ be a table of a data set D, $A = A_1, A_2, ..., A_m$ be all the attributes of T, and $t_i[A_j]$ be the value of attribute A_j of tuple t_i. If $C = C_1, C_2, ..., C_k \subseteq A$, then we denote $T[C] = t[C_1], t[C_2], ..., t[C_k]$ as the projection of t onto the attributes in C.

As defined above, quasi-identifier is a set of attributes that re-identify an individual by a combination of several or all of them with a high probability. The quasi-identifiers can be linked with other data sources to exclusively re-identify a record in the dataset. To make it concise, QI is used to denote a set of all quasi-identifiers in the following context.

A table T satisfies k-anonymity if for every tuple $t \in T$ there exist at least $k - 1$ other tuples $t_{i_1}, t_{i_2}, ..., t_{i_{k-1}} \in T$, such that $t[C] = t_{i_1}[C] = t_{i_2}[C], ..., t_{i_{k-1}}[C]$, for all $C \in QI$.

It is also worth mentioning that a greater value of k leads to a higher data utility loss. Moreover, k-anonymity is vulnerable to homogenous attacks and causes privacy leakage because of the homogeneity of sensitive attributes. For instance, if a data attacker already held background knowledge of "Linda had cancer and was in Table 2.2, it can be inferred that the fourth record of Table 2.2 belongs to Linda.

Table 2.2 An Instance of k-anonymity (k=2)

Job	Gender	Age	Illness	Non-sensitive
Artist	F	15–35	Hay Fever	N.A.
Artist	M	25–40	HIV	N.A.
Educators	M	20–30	Fever	N.A.
Educators	F	10–30	Cancer	N.A.

Table 2.3 An Instance of l-anonymity (k=2, l=2)

Job	Gender	Age	Illness	Other
Engineer	M	40–50	Diabetes	NA
Engineer	M	40–50	Diabetes	NA
Engineer	M	40–50	Cancer	NA
Engineer	M	40–50	Cancer	NA

Motivated by this, Machanavajjhala et al. proposed l-diversity model in 2006 [19]. As its name indicates, l-diversity puts focus on the diversity of sensitive attributes. It requires that at least one sensitive attribute value is different in each anonymous group. Based on this, the maximum probability of a data attacker successfully re-identifying a record is $\frac{1}{l}$. We show an example of l-diversity in Table 2.3, where $k = 2$ and $l = 2$.

Intuitively, l-diversity is an extended version of k-anonymity, which "well-tuned" the diversity of sensitive attributes. There are four representative variants of l-diversity, which are articulated as follows.

- Distinct l-diversity: Similar to k-anonymity, each sensitive attribute has to possess at least l distinct values in each QI group.
- Probabilistic l-diversity: The frequency of a sensitive value in a QI group is at most $\frac{1}{l}$.
- Entropy l-diversity: For every QI group, its entropy is at least $\log l$.
- (c, l)-diversity: The frequency of sensitive values of a QI group is confined in the range defined by c (a real number) and l (in integer).

Unfortunately, l-diversity its own limitations, among which the most serious one is vulnerable to similarity attacks. The rationale behind this is a data attacker can infer the sensitive information of an user based on the sensitive familiarity value and the semantic similarity of each QI group. In an improperly designed situation, l-diversity may leak much more sensitive information than expectation.

This is also the driving force of t-closeness, which is proposed by Li et al. in 2010. T-closeness is extended from k-anonymity and l-diversity but focuses more on data distribution. The difference between the distributions of a QI group and that of the raw data should be within a pre-defined threshold. This brings further privacy protection to the datasets.

Built upon the above three models, several other anonymity and clustering based models are proposed, including (a, k)-anonymous [28], (k, e)-Anonymous [29], and (e, m)-Anonymous [30], etc. To sum up, anonymity and clustering based methods are established on specific attack assumptions, and consequently can not be quantitative analyzed. Moreover, the scalability are not very good due to the consideration of diversity and closeness. This stops them from further proliferation in real-world IoT scenarios.

2.4 Differential Privacy Based Methods

Differential privacy is first proposed by Dwork in 2006. It is a probabilistic model that has some unique features compared to cryptography-based methods and anonymity and clustering based methods. It provides strict privacy preservation under the framework of information theory.

The idea of differential privacy roots in differential attacks. To launch differential attacks, a data attacker may send multiple queries to a statistical database based on his/her background knowledge. This can significantly increase the probability of re-identifying a record.

The idea of differential privacy is to minimize the information gain between queries on two adjacent datasets. The adjacent datasets means two datasets with a single different record, which is also the sensitive record that a data attack is interested in. For numeral data, the most popular solution is adding randomized but controllable noises complying with a specific distribution like Laplace distribution, Gassian distribution, etc.

Definition 2.1 *Differential Privacy:* A random function M satisfies ϵ-differential privacy if for every $D_1 \sim D_2$, and for all outputs $t \in P$ of this randomized function, the following statement holds:

$$P_r[M(D_1)] \leq exp(\epsilon)P_r[M(D_2)], \tag{2.1}$$

in which exp refers to the exponential function. Two data sets D_1 and D_2 are adjacent with at most one different record. ϵ is the privacy protection parameter that controls the degree of difference induced by two neighbouring data sets. A smaller ϵ leads to a stronger privacy guarantee but lower data utility.

ϵ-differential privacy is achieved by injecting random noise whose magnitude is adjustable based on ϵ and the global sensitivity. The definition of global sensitivity is as follows.

Definition 2.2 *Global Sensitivity:* The global sensitivity $S(f)$ of a function f is the maximum absolute difference obtained on the output over all adjacent datasets:

$$S(f) = \max_{D_1 \sim D_2} |f(D_1 - D_2)|. \tag{2.2}$$

Two popular randomized mechanisms are widely deployed to meet the requirements of differential privacy, which are Laplace mechanism and the Exponential mechanism. As mentioned above, adding noise is a mainstream method, which is achievable using Laplace mechanism in numeral cases, which is also the most general case.

Definition 2.3 *Laplace Machanism:* Given a function $f: D \to P$, the mechanism $M: R \to \triangle(R)^n$ adds Laplace distributed noise to the output of f:

$$M(D) = f(D) + V, where V \sim Lap\left(\frac{S(f)}{\epsilon}\right), \tag{2.3}$$

where $Lap\left(\frac{S(f)}{\epsilon}\right)$ has PDF $\frac{1}{2\sigma}exp(\frac{-\epsilon|x|}{\sigma})$, $\sigma = \frac{S(f)}{\epsilon}$ is the scale parameter. The novel algorithm developed in this paper adopts the standard Laplacian mechanism.

Wong et al. [28] found that differential privacy does not match the legal definition of privacy, which is required to protect individually identifiable data, rather than the how much one individual can affect an output as differential privacy provides. As a result, they proposed differential identifiability to provide strong privacy guarantees of differential privacy, while letting policy-makers set parameters based on the established privacy concept of individual identifiability. Following this research line, Zhang et al. [29] analyzed the pros and cons of differential privacy and differential identifiability and proposed a framework called membership privacy. The proposed framework offers a principled approach to developing new privacy notions under which better utility can be achieved than what is possible under differential privacy.

The traditional privacy is a global model for all records within a dataset. Therefore, the privacy protection level is equivalent to all records. That is also the reason why it is named uniform differential privacy or homogenous differential privacy. However, the privacy protection level of each record varies a lot in real-world case. To make this happen, personalized differential privacy, which is also know as heterogenous differential privacy or non-uniform differential privacy, is widely investigated [30]. Besides, Qu et al. has studied this problem for various application scenarios, including edge computing, cyber-physical systems, social netowrks, etc. [31–35].

2.5   Machine Learning and AI Methods

The fast proliferation of machine learning (ML), IoT, and AI drives the emergence of edge intelligence, which brings a lot of benefits to daily life. However, it requires extensive data to train a ML model before any services can be provided. This brings about growing privacy leakage concerns on the data collection and model training.

The good news is that recent research has testified that ML models can also be leveraged as a tool for privacy preservation, especially in IoT scenarios. Nowadays, innovative decentralized learning paradigms emerge, which enables distributive learning task on local data [36]. This type of paradigms has attracted attention from researchers, engineers, and public worldwide.

Usually, distributed training system contains the following main modules: data and model partitioning module, stand-alone optimization module, a communication module, as well as data and model aggregation module. Under this framework, local machines are assigned with different tasks of a ML model using different data.

In this way, privacy preservation on local data is achieved against leakage during transmission and malicious central servers like cloud.

Internet of Things is a widely deployed semi-decentralized computing and network paradigm that enables local task processing on IoT devices. Analogous to distributed computing, IoT can alleviate privacy issues to some extent [37]. However, it is necessary to integrate other advanced technologies to provide proper privacy or security protection. For instance, Gai et al. jointly used blockchain and edge computing for privacy preservation in IoT, especially in smart gird [38]. By leveraging additive secret sharing and edge computing, Ma et al. developed a privacy-preserving mode for face recognition on IoT devices [39]. Besides, based on the Boneh-Goh-Nissim cryptography system, Li et al. carried out a data aggregation scheme with privacy preservation for IoT applications [40]. In addition, Du et al. made efforts to preserve the privacy of training datasets using differential privacy in wireless IoT scenarios. On account of this, machine learning models for privacy preservation are further proposed and applied in practice, among which federated learning and GAN are the most popular ones.

2.5.1 Federated Learning

Federated learning is a innovative distributed learning paradigm booming in recent years. Different from existing distributed learning paradigms, federated learning enables the central server and local IoT devices to maintain the same machine learning model by exchanging model parameters instead of raw data [41]. In this way, local IoT devices use a warm-up model to training local data and then submit local model parameters to the central server for global aggregation [42]. The aggregated global model parameters are then returned to IoT devices to the next round of local training. This process iterates until the performance of the global model meets the requirements.

Several variants of federated learning has emerged, which solves the limitations of classic federated learning paradigm from various perspectives. The first one is asynchronous federated learning, which aggregates the local models in an asynchronous way [43, 44]. This can improve the efficiency while defeating various attacks. The second one is decentralized federated learning using blockchain [45]. The decentralized paradigm removes the central server and uses a consensus algorithm to perform global aggregation [34, 46]. It is robust to a wide range of attacks and privacy risks [47].

However, existing research has shown that federated learning may be subject privacy leakage as well [48]. By analyzing local model parameters, adversaries are able to launch reverse attacks. Now, various new algorithms or techniques are deployed to enhance privacy protection, for example, differential privacy [49], model compression, etc.

2.5.2   Generative Adversarial Network

Generative Adversarial Network (GAN) has playing an increasingly important role in privacy preservation domain. At first, GAN is proposed to generate synthetic images to mimic raw images [50]. This brings natural privacy-preserving features to it. Then, it is extended to preserve the privacy of numeral data as well.

With the development of GAN, researchers found that it cannot provide formal privacy protection to the data, which makes it an empirical solution [51]. Therefore, new research directions emerge to combine GAN and differential privacy in both ways, in particular, differentially private GAN [52] and GAN-enhanced differential privacy [33, 53, 54]. Moreover, GAN has been deployed to provide privacy protection to federated learning by generating privacy-preserving model parameters [55, 56].

New paradigms are emerging every single day. We look forward to more machine learning models can provide privacy preservation or dedicated privacy-preserving machine leaning paradigms in the near future.

References

1. J. Yu, K. Wang, D. Zeng, C. Zhu, S. Guo, Privacy-preserving data aggregation computing in cyber-physical social systems. ACM Trans. Cyber-Phys. Syst. **3**(1), 1–23 (2018)
2. L. Cui, G. Xie, Y. Qu, L. Gao, Y. Yang, Security and privacy in smart cities: challenges and opportunities. IEEE Access **6**, 46 134–46 145 (2018)
3. L. Cui, Y. Qu, L. Gao, G. Xie, S. Yu, Detecting false data attacks using machine learning techniques in smart grid: a survey. J. Netw. Comput. Appl., 102808 (2020)
4. L. Cui, Y. Qu, L. Gao, G. Xie, S. Yu, Blockchain based decentralized privacy preserving in edge computing, in *Privacy-Preserving in Edge Computing*. Springer, pp. 83–109 (2021)
5. L. Cui, Y. Qu, L. Gao, G. Xie, S. Yu, Context-aware privacy preserving in edge computing, in *Privacy-Preserving in Edge Computing*. Springer, pp. 35–63 (2021)
6. L. Cui, Y. Qu, L. Gao, G. Xie, S. Yu, An introduction to edge computing, in *Privacy-Preserving in Edge Computing*. Springer, pp. 1–14 (2021)
7. L. Cui, Y. Qu, L. Gao, G. Xie, S. Yu, Privacy issues in edge computing, in *Privacy-Preserving in Edge Computing*. Springer, pp. 15–34 (2021)
8. Y. Qu, L. Gao, Y. Xiang, Blockchain-driven privacy-preserving machine learning, *Blockchains for Network Security: Principles, Technologies and Applications*, pp. 189–200 (2020)
9. L. Gao, T.H. Luan, B. Gu, Y. Qu, Y. Xiang, *Privacy-Preserving in Edge Computing, ser* (Springer, Wireless Networks, 2021)
10. Y. Qu, M.R. Nosouhi, L. Cui, S. Yu, Personalized privacy protection in big data
11. Y. Qu, M.R. Nosouhi, L. Cui, S. Yu, Existing privacy protection solutions, in *Personalized Privacy Protection in Big Data*. Springer, pp. 5–13 (2021)
12. Y. Qu, M.R. Nosouhi, L. Cui, S. Yu, Future research directions, in *Personalized Privacy Protection in Big Data*. Springer, pp. 131–136 (2021)
13. Y. Qu, M.R. Nosouhi, L. Cui, S. Yu, Leading attacks in privacy protection domain, in *Personalized Privacy Protection in Big Data*. Springer, pp. 15–21 (2021)
14. Y. Qu, M.R. Nosouhi, L. Cui, S. Yu, Personalized privacy protection solutions, in *Personalized Privacy Protection in Big Data*. Springer, pp. 23–130 (2021)
15. A. Sahai, B. Waters, Fuzzy identity-based encryption, in *Annual International Conference on the Theory and Applications of Cryptographic Techniques*. Springer, pp. 457–473 (2005)

16. V. Goyal, O. Pandey, A. Sahai, B. Waters, Attribute-based encryption for fine-grained access control of encrypted data, in *Proceedings of the 13th ACM conference on Computer and communications security*, pp. 89–98 (2006)
17. A. Lewko, B. Waters, Decentralizing attribute-based encryption, in *Annual international conference on the theory and applications of cryptographic techniques*. Springer, pp. 568–588 (2011)
18. D.X. Song, D. Wagner, A. Perrig, Practical techniques for searches on encrypted data, in *Proceeding, IEEE Symposium on Security and Privacy. S&P 2000*. IEEE, pp. 44–55 (2000)
19. R. Curtmola, J. Garay, S. Kamara, R. Ostrovsky, Searchable symmetric encryption: improved definitions and efficient constructions. J. Comput. Secur. **19**(5), 895–934 (2011)
20. C. Gentry, Fully homomorphic encryption using ideal lattices, in *Proceedings of the forty-first annual ACM symposium on Theory of computing*, pp. 169–178 (2009)
21. V. Vaikuntanathan, Computing blindfolded: New developments in fully homomorphic encryption, in *IEEE 52nd Annual Symposium on Foundations of Computer Science*. IEEE, pp. 5–16 (2011)
22. A.C. Yao, Protocols for secure computations, in *23rd annual Symposium on Foundations of Computer Science (sfcs.)* IEEE, pp. 160–164 (1982)
23. Q. Sun, D.R. Simon, Y.-M. Wang, W. Russell, V.N. Padmanabhan, L. Qiu, Statistical identification of encrypted web browsing traffic, in *Proceedings, IEEE Symposium on Security and Privacy*. IEEE, pp. 19–30 (2002)
24. M. Liberatore, B.N. Levine, Inferring the source of encrypted http connections, in *Proceedings of the 13th ACM Conference on Computer and Communications Security*, pp. 255–263 (2006)
25. Y. Zhu, X. Fu, B. Graham, R. Bettati, W. Zhao, Correlation-based traffic analysis attacks on anonymity networks. IEEE Trans. Parallel Distrib. Syst. **21**(7), 954–967 (2009)
26. P. Venkitasubramaniam, T. He, L. Tong, Anonymous networking amidst eavesdroppers. IEEE Trans. Inform. Theory **54**(6), 2770–2784 (2008)
27. S. Yu, G. Zhao, W. Dou, S. James, Predicted packet padding for anonymous web browsing against traffic analysis attacks. IEEE Trans. Inform. Forensics Secur. **7**(4), 1381–1393 (2012)
28. R.C.-W. Wong, J. Li, A.W.-C. Fu, K. Wang, (α, k)-anonymity: an enhanced k-anonymity model for privacy preserving data publishing, in *Proceedings of the 12th ACM SIGKDD International Conference on Knowledge Discovery and Data Mining*. ACM, pp. 754–759 (2006)
29. Q. Zhang, N. Koudas, D. Srivastava, T. Yu, Aggregate query answering on anonymized tables, in *IEEE 23rd International Conference on Data Engineering*. IEEE, pp. 116–125 (2007)
30. J. Li, Y. Tao, X. Xiao, Preservation of proximity privacy in publishing numerical sensitive data, in *Proceedings of the 2008 ACM SIGMOD International Conference on Management of Data*. ACM, pp. 473–486 (2008)
31. Y. Qu, S. Yu, W. Zhou, S. Peng, G. Wang, K. Xiao, Privacy of things: emerging challenges and opportunities in wireless internet of things. IEEE Wirel. Commun. **25**(6), 91–97 (2018)
32. Y. Qu, S. Yu, L. Gao, W. Zhou, S. Peng, A hybrid privacy protection scheme in cyber-physical social networks. IEEE Trans. Comput. Soc. Syst. **5**(3), 773–784 (2018)
33. Y. Qu, S. Yu, J. Zhang, H.T.T. Binh, L. Gao, W. Zhou, Gan-dp: Generative adversarial net driven differentially privacy-preserving big data publishing, in *ICC 2019-2019 IEEE International Conference on Communications (ICC)*. IEEE, pp. 1–6 (2019)
34. Y. Qu, L. Gao, T.H. Luan, Y. Xiang, S. Yu, B. Li, G. Zheng, Decentralized privacy using blockchain-enabled federated learning in fog computing. IEEE Int. Things J. (2020)
35. Y. Qu, L. Cui, S. Yu, W. Zhou, J. Wu, Improving data utility through game theory in personalized differential privacy, in *IEEE International Conference on Communications, ICC 2018, Kansas City, America, May 21–25, 2018*, pp. 1–6 (2018)
36. Y. Sun, J. Liu, J. Wang, Y. Cao, N. Kato, When machine learning meets privacy in 6g: a survey. IEEE Commun. Surv. Tutor. (2020)
37. J. Zhang, B. Chen, Y. Zhao, X. Cheng, F. Hu, Data security and privacy-preserving in edge computing paradigm: survey and open issues. IEEE Access **6**, 18 209–18 237 (2018)
38. K. Gai, Y. Wu, L. Zhu, L. Xu, Y. Zhang, Permissioned blockchain and edge computing empowered privacy-preserving smart grid networks. IEEE Int. Things J. **6**(5), 7992–8004 (2019)

39. Z. Ma, Y. Liu, X. Liu, J. Ma, K. Ren, Lightweight privacy-preserving ensemble classification for face recognition. IEEE Int. Things J. **6**(3), 5778–5790 (2019)
40. X. Li, S. Liu, F. Wu, S. Kumari, J.J. Rodrigues, Privacy preserving data aggregation scheme for mobile edge computing assisted iot applications. IEEE Int. Things J. **6**(3), 4755–4763 (2018)
41. X. Wang, Y. Han, C. Wang, Q. Zhao, X. Chen, M. Chen, In-edge ai: intelligentizing mobile edge computing, caching and communication by federated learning. IEEE Network (2019)
42. J. Konecný H.B. McMahan, F. X. Yu, P. Richtárik, A.T. Suresh, D. Bacon, Federated learning: Strategies for improving communication efficiency CoRR arXiv:1610.05492 (2016)
43. Y. Liu, Y. Qu, C. Xu, Z. Hao, B. Gu, Blockchain-enabled asynchronous federated learning in edge computing. Sensors **21**(10), 3335 (2021)
44. C. Xu, Y. Qu, Y. Xiang, L. Gao, Asynchronous federated learning on heterogeneous devices: a survey, arXiv preprint arXiv:2109.04269 (2021)
45. C. Xu, Y. Qu, P.W. Eklund, Y. Xiang, L. Gao, Bafl: an efficient blockchain-based asynchronous federated learning framework, in *IEEE Symposium on Computers and Communications (ISCC)*. IEEE **2021**, pp. 1–6 (2021)
46. Y. Qu, S.R. Pokhrel, S. Garg, L. Gao, Y. Xiang, A blockchained federated learning framework for cognitive computing in industry 4.0 networks. IEEE Trans. Indus. Inform. **17**(4), 2964–2973 (2020)
47. Y. Wan, Y. Qu, L. Gao, Y. Xiang, Privacy-preserving blockchain-enabled federated learning for b5g-driven edge computing. Comput. Netw., 108671 (2021)
48. G. Xu, H. Li, S. Liu, K. Yang, X. Lin, Verifynet: secure and verifiable federated learning. IEEE Trans. Inform. Forensics Secur. **15**, 911–926 (2019)
49. M. Hao, H. Li, X. Luo, G. Xu, H. Yang, S. Liu, Efficient and privacy-enhanced federated learning for industrial artificial intelligence. IEEE Trans. Indus. Inform. **16**(10), 6532–6542 (2019)
50. I.J. Goodfellow, J. Pouget-Abadie, M. Mirza, B. Xu, D. Warde-Farley, S. Ozair, A.C. Courville, Y. Bengio, Generative adversarial nets, in *Advances in Neural Information Processing Systems 27: Annual Conference on Neural Information Processing Systems 2014, December 8–13 2014, Montreal, Quebec, Canada*, pp. 2672–2680 (2014)
51. S. Ho, Y. Qu, L. Gao, J. Li, Y. Xiang, Generative adversarial nets enhanced continual data release using differential privacy, in *International Conference on Algorithms and Architectures for Parallel Processing*. Springer, pp. 418–426 (2019)
52. G. Ács, L. Melis, C. Castelluccia, E.D. Cristofaro, Differentially private mixture of generative neural networks, in *2017 IEEE International Conference on Data Mining, ICDM 2017, New Orleans, LA, USA, November 18–21, 2017*, pp. 715–720 (2017)
53. Q. Youyang, Z. Jingwen, L. Ruidong, Z. Xiaoning, Z. Xuemeng, Y. Shui, Generative adversarial networks enhanced location privacy in 5g networks. Sci. China Inform. Sci.
54. Y. Qu, S. Yu, W. Zhou, Y. Tian, Gan-driven personalized spatial-temporal private data sharing in cyber-physical social systems. IEEE Trans. Network Sci. Eng. **7**(4), 2576–2586 (2020)
55. L. Cui, Y. Qu, G. Xie, D. Zeng, R. Li, S. Shen, S. Yu, Security and privacy-enhanced federated learning for anomaly detection in iot infrastructures. IEEE Trans. Indus. Inform. (2021)
56. Y. Wan, Y. Qu, L. Gao, Y. Xiang, Differentially privacy-preserving federated learning using wasserstein generative adversarial network, in *IEEE Symposium on Computers and Communications (ISCC)*. IEEE **2021**, pp. 1–6 (2021)

Chapter 3
Decentralized Privacy Protection of IoTs Using Blockchain-Enabled Federated Learning

Internet of Things (IoT), as the extension of cloud computing and a foundation of edge computing, is fast booming due to its abilities to easy the bothersome issues of networks like high latency, congestion, etc. However, the privacy issues attract increasing concerns which result in the "data isolated islands" while dragging down the performances of IoT. Most existing models in this domain hardly consider this and are vulnerable to data falsification attacks. Motivated by this, we carry out a innovative decentralized federated learning built upon blockchain (FL-Block) to address the above issues. In FL-Block, miners train local models using local data and then exchange local model parameters. Then, the consensus algorithm enables the global aggregation in a decentralized way. In this way, FL-Block allows autonomous machine learning that is robust against various privacy and security risks. Moreover, the communicating latency of FL-Block is analyzed and further obtain the optimal block generation rate by comprehensively considering communication, consensus delays, and computation cost. Extensive experimental results show that FL-Block's performances is superior from the perspectives of privacy protection, model convergence efficiency, resistance to data falsification attacks. This chapter is mainly based on our recent research decentralized federated learning [1–5].

3.1 Overview

Internet of Things, which works as a synthesis of cloud computing and edge computing, is experiencing fast popularity because of its proficiency in various aspects, such as mass data collection, network latency and congestion reduction while potentially enabling autonomous systems [6]. IoT has shifted the way of data sharing, content distributing, and other services [7]. Built upon current practice and methodologies,

© The Author(s), under exclusive license to Springer Nature Singapore Pte Ltd. 2022
Y. Qu et al., *Privacy Preservation in IoT: Machine Learning Approaches*,
SpringerBriefs in Computer Science, https://doi.org/10.1007/978-981-19-1797-4_3

19

interdependencies of network entities have been developed in IoT for prospective applications [8]. IoT is also able to manage social interactions of IoT devices with homogeneous objectives, functions. This can be extended to mobility patterns in the virtual community of IoT devices, users, and edge servers deployed in IoTs.

Although diverse conveniences has been carried out, the emerging privacy issues become the bottleneck that stops it from further application. In classic cases, IoT devices are required to upload its data to a trusted central authority like cloud servers for processing [9]. Usually, the shared data contains sensitive information or even exclusive identifiers, such as IP address, location, etc. [10, 11]. That is also the reason why privacy leakage is increasingly severe. At the same time, data utility should not significantly degrade despite the implementation of privacy protection method [12]. Otherwise, the trained machine learning model may have so low performances that it is impracticable in real-world scenarios.

What's more, the communication overhead increases exponentially in IoTs due to the existence of hundreds of millions of IoT devices. The data sharing and uploading process in IoT networks bring extra burden to the systems [13, 14]. Under this circumstance, the guarantee of real-time communication in IoT networks requires further investigation [15].

To address the aforementioned issues, some existing works have made attempts from different angles. Classic privacy protection framework, differential privacy [16] with its extensions [17], is leveraged to deal with the aggregation privacy. However, it is difficult to find a universal optimal trade-off between data utility and privacy protection [12]. Encryption-based methods, such as the work conducted in [18], can preserve privacy in a satisfying degree but is not scalable to big data scenarios. Federated learning [19], which is a new-merging technology, can achieve efficiency communication by exchanging updates between local models and global models [7]. But another concern arises that the central authority may be compromised and poisoning attacks are launched by adversaries. Blockchain is introduced to this scenario to eliminate the trust issues here [20]. But strictly providing privacy protection only on blockchain and efficiency encumber the feasibility of direct applications.

To address the aforementioned issues, we develop a decentralized federated learning paradigm based on blockchain (FL-Block). It can achieve decentralized global aggregation by means of the underlying blockchain architecture. In FL-Block, IoT devices broadcast its local model parameters to other participants. The tailor-made consensus algorithm decides which the winning miner to conduct global aggregation. To meet the constraints of limited resources, only a pointer of model parameters are stored on IoT devices. The global model parameters are stored in a distributed hash table (DHT), which is mapped by the pointers. With a hybrid identity generation, comprehensive verification, access control, and off-chain data storage and retrieve, FL-Block enables decentralized privacy protection while being able to defeat data manipulation attacks and avoiding single-point failure. Furthermore, the incentive mechanism of blockchain can better improve the participant rate and consequently improve the model performance.

The main contributions of this work are summarized as follows.

- **Decentralized Privacy Preservation**: By jointly leveraging federated learning and blockchain, FL-Block achieves decentralized privacy preservation by designing protection protocols. Furthermore, FL-Block offers incentive mechanism to the participants of federated learning to further improve the performance.
- **Against Data Manipulation Attack**: Data manipulation attacks can be defeated with FL-Block because of the existence of blockchain with cross-validation. This enables non-tempering and anomaly detection, which leads to the feature of data manipulation attack proof.
- **High Convergence Efficiency**: The high convergence efficiency is achieved due to the following two reasons. Firstly, federated learning exchanges the model parameters instead of raw data. Secondly, the data manipulation attack can be defeated such that the global model can converge faster.

The remainder of this chapter is organized as follows. In Sect. 3.2, we present current research on relevant research in blockchain, edge computing, and federated learning. We continue to demonstrate decentralized privacy protection model using blockchain-enabled federated learning in Sects. 3.3 and 3.4, respectively. Section 3.5 presents the analysis on efficiency and block generation with and without optimal setting, which is followed by evaluations on real-world data sets in Sect. 3.6. At last, we conclude and summurize this chapter in Sect. 3.7.

3.2 Related Work

In edge computing, the environment is open to access, which poses great challenges from the perspective of privacy protection in real-world applications [21]. Hu et al. [22] devised a novel Identity-based scheme to solve device-to-device and device-to-server communications in edge computing. However, there are potential risks due to the storage of the system's master key in every end device [23]. In addition, the scalability and resultant communication overhead are not taken into consideration. Privacy protection attracts increasing volume of interest from both academia and industry [24, 25]. From the perspective of anonymity, Kang et al. [26] designed a hybrid model that enhances the privacy protection by means of pseudonyms. Employing self-certification, it is possible to manage without the compromise of the system's robustness. To further improve, Tan et al. [27] developed an end device identity authentication protocol of end devices and edge servers without certification, while Wazid et al. [28] proposed another secure authentication and key management scheme to improve key security of the users in IoT scenarios. In [29], Gu et al. proposed a customizable privacy protection model to enable the personalized privacy-preserving data sharing using Markov decision process. This enables the private data sharing between end devices and edge servers, which meets numerous privacy demands. Nevertheless, these three models requires a trusted third party, which is another bottleneck in term of decentrilized privacy protection [30].

Blockchain has a specially designed distributed ledger structure that connects blocks in chronological order. The saved data is shared and maintained by all nodes in a decentralized environment. The leading advantages of blockchain are decentralization, non-tempering, open autonomy, and anonymous traceability [31]. By using blockchain, Jiao et al. [32] proposed a novel scheme to manage resource allocation for edge or cloud computing. Then, Rowan et al. [33] designed a blockchain-enhanced PKI with an inter-end device session key establishment protocol to provide privacy protection to device-to-device communications. Peng et al. [34] presented an onboard network anonymous authentication protocol. The strength of this model is guaranteeing the anonymity of users and efficient authentication, but it fails to detect malicious end devices. In addition, Lu et al. proposed a blockchain-based reputation system and Malik proposed a blockchain-based authentication in [35, 36], respectively. Despite the characteristic of privacy preservation, these two models are incapable of end device announcement, which results in low scalability. To address the limitation of pre-storing lots of anonymous certificates, Lu et al. [18] designed a novel model to achieve conditional privacy. This model requires frequent change of anonymous certificates to avoid linkage attack, which hinders the performance of edge computing in term of efficiency. In [37], blockchain meets distribute hash table (DHT), which decouples validation from state storage.

Federated learning is a potential promising tool, which is proposed as a decentralized learning scheme where local data is distributed to the data owner [38, 39]. Federated learning enables each data owner to have a series of local learning parameters from the local model. Rather than simply sharing the training data, the data owners share their local parameter updates with a trusted third party. Konecny et al. [40] explored the applicability of several current models and thereby devised an extension to address the sparse data issues. In [41], Konecny et al. further present a variant of federated learning which minimizes the communication cost by uploading a lessened number of local parameters. Smith et al. [39] devised a distributed learning model in term of multiple relevant tasks using federated learning. Nishio et al. [42] also gave insights on how to select clients in a resource-limited mobile edge computing scenario. Lu et al. integrated federated learning into the consensus algorithm of blockchain to save the hashrate [43]. Samarakoon et al. [44] devised a learning based intrusion detection based on sample selected extreme learning in edge computing. However, this work did not take sharing resources and device-to-device communication into consideration.

Interested readers are encouraged to explore literature review we have done in recent years [45–49]. Besides, we also summarize several of our book collections to show the frontier of privacy preservation research in edge computing [50–55] and privacy preservation research in edge computing in big data [56–60].

To the best of our knowledge, no existing work has studied the joint use of blockchain and federated learning in the context of efficiently privacy-preserving communication in edge computing scenario.

3.3 Architecture of Blockchain-Enabled Federated Learning

In this section, we present the architecture of the blockchain-enabled federated learning (FL-Block). We use v_i and m_j to denote different end devices and miners, respectively. Then ds_k is used to present different data samples. In addition, we use e_l to differentiate the global model update iterations, which is also referred to as epochs.

3.3.1 Federated Learning in FL-Block

In FL-Block, the federated learning is achieved by a cluster of end devices $V = \{1, 2, ..., N_V \in V\}$ where $|V| = N_V$. The i-th device V_i possesses a set of data samples DS_i where $|DS_i| = N_i$. V_i trains its local data and the local model updates of the device V_i is uploaded to its associated miner M_i instead of a trusted third party. The miner M_i is selected from a set of miners $M = \{1, 2, ..., N_M\}$ randomly, where $|M| = N_M$. $M = V$ could be satisfied when the miners M_i are physically identical to the end devices. Otherwise, we will have $M \neq DS$. Moving on, the total number of v_i of the local model updates are verified and shared among all possible miners. Finally, the aggregated global model updates are downloaded from each miner to its corresponding device.

To further improve, the decentralized model training concentrates on solving a linear regression problem on a series of parallel data samples $DS = \cup_{i=1}^{N_V} DS_i$ where $|DS| = N_S$. The k-th data sample $ds_k \in DS$ is given as $s_k = \{x_k, y_k\}$ for a d-dimensional column vector $x_k \in R^d$ as well as a scalar value $y_k \in R$. The objective of the linear regression problem is to minimize a pre-defined loss function $f(\omega)$ with respect to a d-dimensional column vector $\omega \in R^d$, which is regarded as a global weight. To simplify this process, the pre-defined loss function $f(\omega)$ is selected as the mean squared error (MSE) in the following context.

$$f(\omega) = \frac{1}{N_D S} \sum_{i=1}^{N_V} \sum_{s_k \in S_i} f_k(\omega), \qquad (3.1)$$

where $f_k(\omega)(x_k^T \omega - y_k)^2 / 2$ and the notation $(.)^T$ denotes the vector transpose operation. This could be easily extended to various loss functions under diverse neural network models with minor operations.

For purpose of solving the aforementioned issues, the learning model of the end device V_i is trained locally with the data sample set DS_i using a stochastic variance reduced gradient (SVRG) algorithm, and all local model updates of the devices are aggregated using a distributed approximate Newton (DANE) method, which will be trained to generate the global model update.

To furtehr improve this model, the ceiling of the global model is set to L epochs. For each epoch, the local model of end device V_i is updated with the number N_i of iterations. Therefore, we can have the local weight $\omega_i^{(t,l)} \in R^d$ of the end device V_i at the $t_t h$ local iteration of the l-th epoch as

$$
\begin{aligned}
\omega_i^{(t,l)} = \omega_i^{(t-1,l)} \\
-\frac{\beta}{N_i}\left\{\left[\nabla f_k(\omega_i^{(t-1,l)}) - \nabla f_k(\omega^{(l)})\right] + \nabla f(\omega^{(l)})\right\},
\end{aligned}
\tag{3.2}
$$

where $\beta > 0$ is defined as a step size, $\omega^{(l)}$ is the global weight at the l-th epoch, and $\nabla f(\omega^{(l)}) = 1/N_{DS}. \sum_{i=1}^{N_V}\sum_{s_k \in S_i} \nabla f_k(\omega^{(l)})$ is derived from Eq. 3.1. We use $\omega^{(l)}$ to represent the local weight when the last local iteration of the l-th epoch is finished, i.e., $\omega_i^{(l)} = \omega_i^{(N_i,l)}$. Built upon this, the update of the global weight $\omega^{(l)}$ is formulated as

$$
\omega^{(l)} = \omega^{(l-1)} + \sum_{i=1}^{N_V} \frac{N_i}{N_{DS}}\left(\omega_i^{(l)} - \omega_i^{(l-1)}\right).
\tag{3.3}
$$

The iteration process of updating the local and global weights will continue until the constraint of global weight $\omega^{(l)}$ satisfies $|\omega^{(L)} - \omega^{(L-1)}| \le \epsilon$ is satisfies, where $\epsilon > 0$ is a small positive constant.

In the classic federated learning settings, at the l-th epoch, the end device V_i is supposed to upload its local model update $\left(\omega_i^{(l)}, \{\nabla f_k(\omega^{(l)})\}_{s_k \in S_i}\right)$ to the edge servers, with the model update size δ_m that is identically specified for each end device. The global model updates $\left(\omega^{(l)}, \nabla f(\omega^{(l)})\right)$ with the same size δ_m are computed by the edge servers, which will be downloaded to all end devices after processing. In FL-Block, the edge server entity is substituted with a blockchain network which is detailed discussed in the following subsection.

3.3.2 Blockchain in FL-Block

Regarding the blockchain network settings of FL-Block, the generated blocks and the cross verification by the miners M are devised to upload truthful data of the local model updates. This is achieved by developing a specially designed distributed ledger. The protocols of the distributed ledger are discussed in the next section in detail. Each block in the distributed ledger contains body and header sectors. In the classic blockchain structure, the body sector usually contains a set of verified transactions. In FL-Block, the body stores the local model updates of the devices in V, i.e., $\left(\omega_i^{(l)}, \{\nabla f_k(\omega^{(l)})\}_{s_k \in S_i}\right)$ for the device V_i at the l-th epoch, as well as its local computation time $T_{\{local,i\}}^{(l)}$. Extended from the classic blockchain structure, the header sector stores the information of a pointer to the previous block, block generation rate λ, and the output value of the consensus algorithm, namely, the nonce. In this context, the proof of work (PoW) is used as the instance consensus algorithm, while all other consensus algorithms can be extended into this scenario. In order to store the local model updates of all the end devices, the size of each block is defined as $h + \delta_m N_V$, where h and δ_m denote the header size and model update size, respectively.

Each miner is assigned with a candidate block that contains local model updates information from its binding end devices or other miners, in the order of arrival. The writing procedure will continue until the block size is fully-filled or a maximum waiting time T_{wait} is reached from the beginning of each epoch. To make it simplified, we use a sufficiently long T_{wait} so that every block is filled with local model updates of all end devices.

In the next stage, the miner keeps generating a random nonce until it becomes smaller than a target value using PoW. Once one of the miners M_1 works out the nonce, its corresponding candidate block is allowed to be generated as a new block. Intuitively, the block generation rate λ can be carefully controlled by modifying the difficulty of the deployed PoW consensus algorithm.

Furthermore, the newly-generated block is shared to all other miners for the purpose of synchronizing all existing distributed ledgers. To achieve this, all the other miners who received the newly-generated block are enforced to quit the calculation process. Then, the generated block is linked to the local ledgers maintained by the miners. However, there is a possibility of forking when another miner M_2 happens to generate a candidate block within the sharing delay of the eligible block. Some miners may falsely link the ineligible block to their local ledgers. In FL-Block, forking will mislead the end devices linking the ineligible block to train a poisoned global update and subsequently generate offtrack local updates in the next iteration.

The value of forking frequency is positively correlated with both the blockchain generate rate λ and the block sharing delay. The mitigation of forking will lead to an extra delay, which will be discussed in Sect. 3.5.

Apart from the above operation for local model updates uploading, the blockchain also provides proper rewards for data samples to the end devices, and for the verification operation to the miners, which is referred to as data reward and mining reward, respectively. The data reward of the end devices V_i is received from its corresponding miner, and the amount of the reward is linearly proportional to the size of the data sample N_i. After a block is generated by the miner M_j, the mining reward is earned from the blockchain network. Similarly, the amount of mining reward is linearly proportional to the aggregated data sample size of its all binding end devices, namely, $\sum_{i=1}^{N_{M_j}}$, where N_{M_j} denotes the number of end devices binding with the miner M_j. This will provide the incentive to the miners so that they will process more local model updates while compensating their expenditure for the data reward.

However, there is a side effect in this rewarding system. There might be malicious end devices deceiving the miners by poisoning their actual sample sizes. In this case, the miners are supposed to verify eligible local updates before storing the local model updates in their associated candidate blocks. In this context, the simplified verification operation is conducted by comparing the sample size N_i with its corresponding

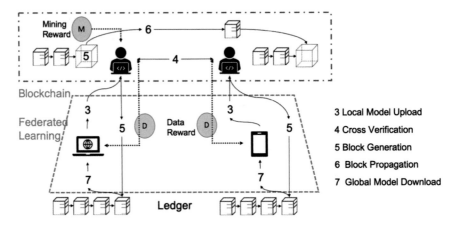

Fig. 3.1 Decentralized federated learning based on blockchain for IoTs

computation time $T^{(}_{\{local,i\}}l)$ that is assumed to be reliable, following the proof of elapsed time under Intel's SGX technology.

As shown in Fig. 3.1, the FL-Block operation of the end device V_i at the l-th epoch can be explained by eight procedures as follows.

Algorithm 1 Block-Enabled Federated Learning Algorithm

Input: IoT devices V_i, Edge servers, Miners M_i;
Output: Optimized Global Model;
1: Initialization;
2: **while** New data sample DS is generated **do**
3:      Local model update by V_i;
4:      Local model upload to edge servers;
5:      Cross verification by M_i;
6:      Block generation by the winning miner;
7:      **if** M_i finds the nonce of the new block b_i **then**
8:          Link b_i to the blockchain
9:          Remove all other relevant blocks $\hat{b}_i$;
10:      **end if**
11:      Block propagation;
12:      Global model update by edge servers;
13:      Global model download to V_i;
14: **end while**
15: Output the eligible global model.

In the initialization part ($l = 1$), the initial parameters are uniformly randomly chosen from predefined ranges of the local and global weights $\{\omega_i^{(0)}, \omega^{(0)} \in (0, \omega_{max}]\}$ for a constant ω_{max}, and the global gradient $\nabla f(\omega^{(0)}) \in (0, 1]$.

In local model update phase, the end device V_i computes Eq. 3.2 with the number N_i of iterations.

In local model upload phase, the end device V_i uniformly randomly associates with the miner M_i. If $M = V$, miner M_i is selected from M V_i. The end device uploads the local model updates $\left(\omega_i^{(l)}, \{\nabla f_k(\omega^{(l)})\}_{s_k \in S_i} \right)$ and the corresponding local computation time $T_{\{local,i\}}^{(l)}$ to the associated miner.

In cross verification phase, the miners broadcast the local model updates obtained from their associated end devices. At the same time, the miners verify the received local model updates from their associated devices or the other miners in the order of arrival. The truthfulness of the local model updates are validated, if the local computation time $T_{\{local,i\}}^{(l)}$ is proportional to the data sample size N_i. The verified local model updates are recorded in the miner's candidate block, until its reaching the block size ($h + \delta_m N_V$) or the maximum waiting time T_{wait}.

In block generation phase, each miner starts running the PoW until either it finds the nonce or it receives a generated block from another miner.

In block propagation phase, denoting as $M_{\hat{o}} \in M$ the miner who first finds the nonce. Its candidate block is generated as a new block that is broadcasted to all miners. In order to avoid forking, an acknowledgment (ACK) signal is transmitted when each miner detects no forking event. Each miner waits until receiving ACK signals of all miners. Otherwise, the operation goes back and iterate from phase 2.

In global model download phase, the end device V_i downloads the generated block from its associated miner M_i.

In global model update phase, the end device V_i locally computes the global model update in Eq. 3.3 by using the aggregate local model updates stored in the generated block.

The one-epoch procedure continues until the global weight $\omega^{(L)}$ satisfies $\omega^{(L)} - \omega^{(L-1)} \leq \epsilon$. The centralized FL structure is vulnerable to the malfunction of edge server, which distorts the global model updates of all end devices. Compared to this, each end device in FL-Block locally computes its global model update, which is thus more robust to the malfunction of the miners that replace the fog server entity.

3.4 Decentralized Privacy Mechanism Based on FL-Block

This section presents the proposed decentralized privacy mechanism in IoTs built upon FL-Block. In Fig. 3.2, the network of IoT devices is referred as distributed hash table (*DHT*). End devices communicate with edge servers through various provided services.

3.4.1 Blocks Establishment

To establish blocks, we orderly define hybrid identity, memory of blockchain, policy, and auxiliary functions.

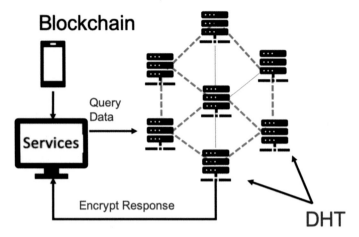

Fig. 3.2 Decentralized privacy in IoTs based on FL-Block

3.4.1.1 Hybrid Identity

The traditional blockchain identity uses a naive anonymous mechanism. By using a public key, every end device can generate an unlimited number of pseudo-identities if necessary. To extend it to our model, we devise a novel *hybrid identity*. This hybrid identity is a form of personalization of identities. When it is shared to different end devices or fog servers, the owner has full access to it while the other parties have limited access according to specific requirements.

In Pro. 1, we illustrate an instance of a sole end device identity owner (u_o) and two guest end devices, namely, identity recipients (u_g). The identity contains generating key-pairs for the sole owner and two guests. In addition, a symmetric key is required to both encrypt and decrypt the data. In this way, this data is restricted to the other end devices in fog computing. The hybrid identity is defined by a 4-tuple as Eq. 3.4.

$$\text{Hybrid}_{u,s}^{(public)} = (\text{pk}_{sig_1}^{u_o,u_{g_1}}, \text{pk}_{sig_2}^{u_o,u_{g_2}}, \text{pk}_{sig_1}^{u_{g_1},u_o}, \text{pk}_{sig_2}^{u_{g_2},u_o}) \tag{3.4}$$

Analogically, the integrated hybrid identity will be a 10-tuple, but we use $i\,|\,i = 1, 2$ as an index to simplify it to a 5-tuple as Eq. 3.5.

$$\text{Hybrid}_{u,s} = (\text{pk}_{sig_i}^{u_o,u_{g_i}}, \text{sk}_{sig_i}^{u_o,u_{g_i}}, \text{pk}_{sig_i}^{u_{g_i},u_o}, \text{sk}_{sig_i}^{u_{g_i},u_o}, \text{sk}_{enc_i}^{u_o,u_{g_i}}) \tag{3.5}$$

3.4.1.2 Memory of Blockchain and Policy

In term of Blockchain memory, let BM be the memory space. We have $BM :$ $0, 1^{256} \rightarrow 0, 1^N$ where $N >> 256$ and it is sufficient to store large data files. Blockchain is like an account book containing a list of model updates with a timestamp. The first two outputs in a model update encode the 256-bit memory address pointer along with some auxiliary meta-data. The other outputs are leveraged to build the serialized document. If $L[k]$ is queried, the model update with the up-to-date timestamp will be returned. This setting allows inserting, delete, and update operation.

We define the policy P_u as a series of permissions that an end device v can gain from a specific service. For instance, if v needs to read, update, and delete a dataset, then $P_v = read, update, delete$. Any data could be safely stored as service will not break the protocols and label the data incorrectly. In addition, the service can easily observe the anomaly of end devices since all changes are visible.

3.4.1.3 Accessional Functions

Two accessional functions are necessary in this case, including Parse(x) and Verify(pk_{sig}^k, x_p). Parse(x) continuously passes the arguments to a specific transaction. Verify(pk_{sig}^k, x_p) help verify the permissions of end devices. The protocol of verify function shown in Pro. 2.

3.4.2 Blockchain Protocols Design

In this subsection, we show more detailed protocols of the proposed blockchain model. When a model update A_{access} is recorded, Pro. 3 is executed by the nodes inside the network. Analogically, when a model update A_{data} is recorded, Pro. 4 is executed by the nodes.

As A_{access} is used to conduct access control management, it can change the permissions of end devices granted by the service. This is accomplished by sending the policy $P_{u_o, u_{g_i}}$. If the service wants to revoke all the access permissions, the policy will be $P_{u_o, u_{g_i}}(\emptyset)$. If it is the first time to send a A_{access} with a new hybrid identity, the A_{access} will be recorded as an end device signing to a service.

Analogously, A_{data} will manage the data manipulation operations such as read, write, update, or delete. With the assistance of the Verify() function, only the service of fog severs or the permitted end devices can access the differentially private data. In Pro. 4, we access the distributed hash table like a normal hash table. In real-world scenarios, these instructions bring about some off-chain network messages which are being sent to the distributed hash table.

3.4.3 Discussion on Decentralized Privacy Protection Using Blockchain

We have some following assumptions in this work. The first one is on the blockchain, which is tamper-proof. This requires a large enough network to avoid untrusted end devices taking over it. The second one is that end devices can properly secure their keys during the operations, for example, using some secure services from fog servers. From here, we will illustrate how the system can protect adversaries from tampering the data stored in a blockchain network. In this scenario, we take less consideration on adversaries modifying protocols or steal personal sensitive information.

In this model, only the services have full control over the sensitive data. An adversary can hardly pretend to be an end device or corrupt the whole network since the blockchain is fully decentralized. In addition, digital signatures are required for model updates. Therefore, we hold that adversaries are not able to fabricate digital signatures or take control of the majority of the network (over 50%). Furthermore, an adversary cannot poison the data because it is stored off-chain rather on the public ledger. There are only pointers information encrypted with a hash function inside a public ledger.

Even if we consider the case that an adversary controls one or some of the nodes in the DHT network, the adversary cannot learn anything about the data. The rationale behind this is that the data is encrypted with keys that no other nodes have access. The worst case is that the adversary gains the authority and compromise a few local copies of the data, the system can still recover it since there are abundant replications distributed across the whole network.

Last but not least, the hybrid identity mechanism ensures that there is only a tiny probability that the data is poisoned because this requires the acquisition of both signing key and encryption-decryption key. If the adversaries happen to steal one of the keys, the sensitive data is still safe. In practice, we can also personalize the hybrid identity so that the compromization is restricted for the adversaries. A good instance would be different keys for a certain volume of records.

3.5 System Analysis

In this section, we first analyze poisoning attacks. Then, we consider a reference end device $V_o \in V$ that is randomly selected. The objective is to derive the optimal block generation rate $\lambda*$, which minimizes the learning completion latency T_o. In this work, we define T_o as the total elapsed time during L epochs at the end device V_o. This latency is linearly proportional to the l-th epoch latency, i.e., $T_o = \sum_{l=1}^{L} T_o^{(l)}$. Thus, we hereafter focus only on $T_o^{(l)}$ without loss of generality.

3.5.1 Poisoning Attacks and Defence

Beyond the naive approach that an adversary train the local model based on specially-designed falsified data, the enhanced attack is referred as model substitution, in which the adversary tries to replace the global model GM^{t+1} with a malicious model MM, which is shown in Eq. 3.6.

$$G_{t+1} = GM^t + \frac{r}{n} \sum_{j=1}^{m} \left(LM_j^{t+1} - GM^t \right)$$

$$\Rightarrow MM = GM^t + \frac{r}{n} \sum_{j=1}^{m} \left(LM_j^{t+1} - GM^t \right)$$

$$(3.6)$$

in which t is the current time slot, r is the learning rate, LM is a local model, n is number the total edge devices, and m is the number participants.

Since non-IID data is used in this case, each local model may be quite different from the global model. With the convergence of the global model, the deviations cancel out bit by bit, which can be denoted as $\sum_{j=1}^{m-1}(LM_j^{t+1} - GM^t) \to 0$. Based on this, an adversary may upload a model as

$$LM_m^{t+1} = \frac{n}{r} MM - \left(\frac{n}{r} - 1 \right) GM^t - \sum_{j=1}^{m-1} \left(LM_j^{t+1} - GM^t \right)$$

$$\simeq \frac{n}{r} \left(MM - GM^t \right) + GM^t.$$

$$(3.7)$$

This enhanced attack increases the weights of the malicious model MM by $\eta = n/r$ to guarantee the replacement of the global model GM by MM. This attack functions better when the convergence is nearly reached. In addition, if the adversary is blind to the value of n and r, he/she can simply increase the value of η in every iteration. Although scaling by $\eta \leq n/r$ only partially replace the GM, the attack still functions well.

3.5.2 Single-Epoch FL-Block Latency Model

Following the FL-Block operation in Sect. 3.3.2, the end device V_o's l-th epoch latency $T_o^{(l)}$ is not only determined by communication and block generation delays but also determined by the computation cost by federated learning.

To begin with, the computation costs are caused by phase 2 and phase 8 defined in Sect. 3.3.2. Let δ_d denote the size of a single data sample that is given identically for all data samples. δ_d/f_c is required to process δ_d with the clock speed f_c. We formulate $T_{\{local,i\}}^{(l)}$ (local model updating delay) in phase 2 as $T_{\{local,i\}}^{(l)} = \delta_d N_o/f_c$. Analogically, we define the global model updating delay $T_{\{global,o\}}^{(l)}$ in phase 8 as $T_{\{global,o\}}^{(l)} = \delta_m N_V/f_c$.

Then, we discuss the communication delays existed in phase 3 and phase 7 between the miners and corresponding end devices. With additive white Gaussian noise (AWGN) channels, we evaluate the achievable rate using Shannon

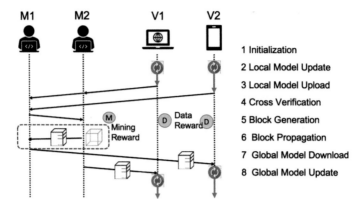

Fig. 3.3 Single-Epoch operation without forking

capacity. In term of phase 3, the local model uploading delay $T^{(l)}_{\{up,o\}}$ is formulated as $T^{(l)}_{\{up,o\}} = \delta_m / \left[W_{up} \log_2(1 + \lambda_{\{up,o\}}) \right]$, where W_{up} is known as the uplink bandwidth allocation for each end device while $\lambda_{\{up,o\}}$ is the received signal-to-noise ratio (SNR) of miner M_o. Similarly, in phase 7, we define the communication delay of global model. The downloading delay $T^{(l)}_{\{vn,o\}}$ is described as $T^{(l)}_{\{vn,o\}} = h + \delta_m N_V / \left[W_{dn} \log_2(1 + \lambda_{\{dn,o\}}) \right]$, where W_{dn} is the downlink bandwidth allocation for each device and $\lambda_{\{dn,o\}}$ is the received signal-to-noise ratio of device V_o (Fig. 3.3).

In addition, the communication delay also exists when miners are communicating in the blockchain network, which is part of phase 4 and phase 5. Since the cross-verification time is insignificant compared with the communication costs, the cross-verification delay $T^{(l)}_{\{cross,o\}}$ in phase 4 is defined under frequency division multiple access (FDMA) as

$$T^{(l)}_{\{cross,o\}} = \max \left\{ T_{wait} - \left(T^{(l)}_{\{local,i\}} + T^{(l)}_{\{up,o\}} \right), \\ \sum_{M_j \in M \setminus M_o} \delta_m N_{M_j} / \left[W_m \log_2(1 + \lambda_{oj}) \right] \right\}, \tag{3.8}$$

where we use W_m to denote the bandwidth allocation per each miner link and λ_{oj} is the received SNR from the miner M_o to the miner M_j. Similarly, $M_{\hat{o}} \in M$ is regarded as the miner who first finds the nonce. Then, the total block propagation delay $T^{(l)}_{\{bp,\hat{o}\}}$ in phase 6 is formulated as $T^{(l)}_{\{bp,\hat{o}\}} = \max_{M_j \in M \setminus M_{\hat{o}}} \left\{ t^{(l)}_{\{bp,j\}} \right\}$ under frequency division multiple access. The term $\left\{ t^{(l)}_{\{bp,j\}} \right\} = (h = \delta_m N_V) / \left[W_m \log_2(1 + \lambda_{\hat{o}j}) \right]$ represents the block sharing delay from the mining winner $M_{\hat{o}}$ to $M_j \in M \setminus M_{\hat{o}}$, and $\lambda_{\hat{o}j}$ is received SNR from the miner $M_{\hat{o}}$ to the miner M_j.

The last considered delay is the block generation delay in phase 5. The miner $M_j \in M$'s block generation delay $T_{\{bg,j\}}^{(l)}$ complies with an exponential distribution with a mean value of $1/\lambda$. The block generation delay $T_{\{bg,\hat{o}\}}^{(l)}$ of the mining winner $M_{\hat{o}}$ is given as the delay of interest. Based on this, the latency $T_o^{(L)}$ of the l-th epoch is described as

$$
T_o^{(l)} = N_{folk}^{(l)} \left(T_{\{local,o\}}^{(l)} + T_{\{up,o\}}^{(l)} + T_{\{cross,o\}}^{(l)} + T_{\{bg,\hat{o}\}}^{(l)} \right.
$$
$$
\left. + T_{\{bp,\hat{o}\}}^{(l)} \right) + T_{\{vn,o\}}^{(l)} + T_{\{global,o\}}^{(l)},
$$

(3.9)

in which the $N_{folk}^{(l)}$ represents forking occurrences' number in the l-th epoch. The forking occurrences complies with a geometric distribution where mean is $1/(1 - p_{folk}^{(l)})$ and the forking probability is $p_{folk}^{(l)}$. Extended from phase 6, the forking probability is defined as

$$
p_{folk}^{(l)} = 1 - \prod_{M_j \in M \setminus M_{\hat{o}}} \Pr\left(t_j^{(l)} - t_{\hat{o}}^{(l)} > t_{bp,j}^{(l)} \right),
$$

(3.10)

where the term $t_j^{(l)} = T_{\{local,j\}}^{(l)} + T_{\{up,j\}}^{(l)} + T_{\{cross,j\}}^{(l)} + T_{\{bg,j\}}^{(l)}$ is the cumulated delay until the miner M_j generates a block (Fig. 3.4).

Fig. 3.4 Single-Epoch operation with forking

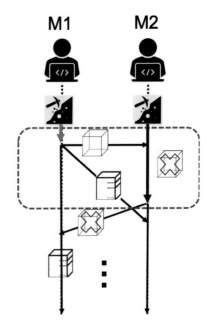

3.5.3 Optimal Generation Rate of Blocks

In this work, we target on deriving the optimal block generation rate λ^*, which can minimize the latency averaged over the consensus algorithm of the end device V_O's l-th epoch by means of the single-epoch latency expression in Eq. 3.9. The consensus process has an impact on three indexes, including the block sharing delay $T^{(l)}_{\{bp,\hat{o}\}}$, the block generation delay $T^{(l)}_{\{bg,\hat{o}\}}$, and the number $N^{(l)}_{folk}$ fork of forking occurrences. These three entities are inter-dependent because of the existence of mining winner M_o. Solving this requires to compare the cumulated delays for all miners and their associated end devicesfog server under their asynchronous operations that complicate the optimization.

To eliminate the above mentioned difficulty, all miners are considered to start their consensus processes synchronously by modifying T_{wait}, with which we further derive $T^{(l)}_{\{cross,o\}} = T_{wait} - (T^{(l)}_{\{local,o\}} + T^{(l)}_{\{up,o\}})$. Under these circumstances, if the miners complete the cross verification phase earlier, they are supposed to wait until T_{wait}, which provides lower bound of the performance, for instance, latency upper bound. Through exact operations using the synchronous approximation, the optimal block generation rate λ^* could be derived in a closed-form as follows.

Proposition 1 *With the PoW synchronous approximation, such as $T^{(l)}_{\{cross,o\}} = T_{wait} - (T^{(l)}_{\{local,o\}} + T^{(l)}_{\{up,o\}})$, the block generation rate λ^* minimizing the l-th epoch latency $E[t^{(l)}_o]$ averaged over the PoW process is given by*

$$\lambda^* \approx 2 \times \left(T^{(l)}_{\{bp,\hat{o}\}} \left[1 + \sqrt{1 + 4N_M \times \left(1 + T_{wait}/T^{(l)}_{\{bp,\hat{o}\}}\right)} \right] \right). \tag{3.11}$$

Next, we apply both the mean $1/(1 - p^{(l)}_{folk})$ of the geometrically distributed $N^{(l)}_{folk}$ and the synchronous consensus approximation to Eq. 3.9 and thus have

$$E[t^{(l)}_o] \approx \left(T_{wait} + E[T^{(l)}_{\{bg,\hat{o}\}}] \right) \Big/ \left(1/(1 - p^{(l)}_{folk}) \right)$$
$$+ T^{(l)}_{\{vn,o\}} + T^{(l)}_{\{global,o\}} \tag{3.12}$$

There are some constant delays, including T_{wait}, $T^{(l)}_{\{vn,o\}}$, and $T^{(l)}_{\{global,o\}}$, which are given in Sect. 3.3.1. Based on these, we further derive the remainder as below.

For the probability $p_{folk}^{(l)}$, using Eq. 3.10 with $t_j^{(l)} - t_{\hat{o}}^{(l)} = t_{\{bg,j\}}^{(l)} - t_{\{bg,\hat{o}\}}^{(l)}$ under the synchronous approximation, we obtain the fork as

$$p_{folk}^{(l)} = 1 - \text{EXP}\left(\lambda \sum_{M_j \in M \setminus M_{\hat{o}}} T_{\{bp,j\}}^{(l)} \right), \tag{3.13}$$

where $T_{\{bp,j\}}^{(l)}$ is a given constant delay. Then, by jointly using $T_{\{bg,\hat{o}\}}^{(l)}$ and the complementary cumulative distribution function (CCDF) of $T_{\{bg,j\}}^{(l)}$ which is the exponentially distributed, CCDF of $T_{\{bg,\hat{o}\}}^{(l)}$ is formulated as

$$\Pr\left(T_{\{bg,\hat{o}\}}^{(l)} > x \right) = \prod_{j=1}^{N_M} \Pr\left(T_{\{bg,j\}}^{(l)} > x \right) = \text{EXP}\left(-\lambda N_M x \right) \tag{3.14}$$

Moreover, we apply the total probability theorem, which yields $E[T_{\{bg,\hat{o}\}}^{(l)}] = 1/(\lambda N_M)$. The final step is to combine all these terms and re-formulize Eq. 3.15 as

$$E[t_o^{(l)}] \approx \left(T_{wait} + 1/(\lambda N_M) \right) \text{EXP}\left(\lambda \sum_{M_j \in M \setminus M_{\hat{o}}} T_{\{bp,j\}}^{(l)} \right)$$
$$+ T_{\{vn,o\}}^{(l)} + T_{\{global,o\}}^{(l)}, \tag{3.15}$$

which is convex with respect to λ. Therefore, the optimal block generation rate λ^* is derived from the first-order necessary condition.

The accuracy of the above result with the synchronous approximation is validated by comparing the simulated λ^* without the approximation in the following section.

3.6 Performance Evaluation

In this section, we testify the superiority of the proposed FL-Block model in term of privacy protection and efficiency. We simulate a fog computing environment and create a blockchain network. Based on this, we conduct experiments on machine learning tasks using real-world data sets of large scale.

3.6.1 Simulation Environment Description

The simulated fog computing environment is implemented on the cellular network of an urban microcell. It is consisted of 3 fog servers, and $N_v = 50$ end devices, on a single workstation. The three fog servers are located at the center of the cell with a radius of 3 km, while the end devices are distributed with a normal distribution in this cell.

To model the wireless communications, we introduce the LTE networks with a popular urban channel model, which is defined in the ITU-R M.2135-1 Micro NLOS model of a hexagonal cell layout [61]. The transmission power and antenna gain of the fog server and end devices are set to be 30 dBm and 0 dBi, respectively, for simplicity. Carrier frequency is set to 2.5 GHz, and the antenna heights of the fog servers and end devices are set to 11 m and 1 m, correspondingly. We use a practical bandwidth limitation, namely, 20 RBs, which corresponds to a bandwidth of 1.8 MHz. This is assigned to an end device in each time slot of 0.5 ms. We employ a throughput model based on the Shannon capacity with a certain loss used in [62], in which $\Delta = 1.6$ and $\rho_{max} = 4.8$. With these initialized settings, the mean and maximum throughputs of client θ_k are 1.4 Mb/s and 8.6 Mb/s, which are practical and feasible in LTE networks. The details of the simulation results are summarized in Table. 3.1 for clarity.

From the aforementioned model, we obtain the throughput and consider it as the average throughput. This throughput is leveraged to derive t in the end device selection phase. We assume the network environment is stable and thereby the federated learning process will not be impacted by network connectivity issues. In this way, we can regard the average throughput as stable outputs. To better simulate the real-world scenarios, we take a small variation of short-term throughput at the scheduled update and upload into consideration. The throughput will be sampled from the Gaussian distribution when the models' updates are shared between end devices and fog servers. The Gaussian distribution is defined by the average throughput and its $r\%$ value as the mean and standard deviation, respectively.

Table 3.1 Simulation parameters overview

Item	Value/description
LTE network	ITU-R M.2135-1 Micro NLOS
Transmission power	30 dBm
Antenna gain	0 dBm
Antenna heights of the fog servers	11 m
Antenna heights of the end devices	1 m
Bandwidth	1.8 MHz
Time slot	0.5 ms
Δ	1.6
ρ_{max}	4.8
Mean throughputs	1.4 Mb/s
Maximum throughputs	8.6 Mb/s

3.6.2 Global Models and Corresponding Updates

In order to guarantee the output's accuracy, the **IID** setting is considered, in which each end device will randomly sample a specific amount of data from the raw dataset.

We implement a classic CNN as the global model for all tasks. In particular, FL-Block contains six 3*3 convolution layers, including 2*32, 2*64, and 2*128 channels. In each of the channel, it is activated by ReLU and batch normalized, while every pair of them is followed by 2*2 max pooling. In additions, the six channels are followed by three fully-connected layers, including 382 and 192 units with ReLU activation and another 10 units activated by soft-max. In this case, the model will approximately have 3.6 million parameters for Fashion-MINIST and 4.6 million parameters for CIFAR-10. The size of D_m are 14.4 mb and 18.3 mb, respectively while the data type is set to be 32-bit float. Although some other deep models will have better performances, they are not the focus of this model and is not considered.

The hyper-parameters of updating global models are initialised as follows. The mini-batch size is 40. The number of epochs in each round is 40. The initial learning rate of stochastic gradient descent updates is 0.30. The learning rate decay is 0.95. We simply model the computation capability of each end device as how many data samples it could train to further update the global model. There might be slight fluctuation due to some other tasks on this end device. We randomly decide the average capability of each end device from the interval [10, 100], which will be used in the client selection phase. Consequently, the update time in client selection phase averagely ranges from 5 to 500 s. The computation capability depends on the Gaussian distribution in both the scheduled update phase and the upload phase. The Gaussian distribution is defined by the average throughput and its $r\%$ value as the mean and standard deviation, respectively. The range is considered to be reasonable since the workstation needs 4 s considering one single update with one single GPU. They may require longer update time up to 100 times if the mobile devices have a weaker computation power. Empirically, we set T_{final} to 400 s and T_{round} to 3 s.

In Tables 3.2 and 3.3, we illustrate the evaluation results with the *IID* settings in terms of ToA and accuracy. Each method is executed for 10 times and the ToA and accuracy scores are measured from the averaging perspective. In term of ToA, FL-Block has excellent performances on both tasks. The architecture is sufficient to show the efficiency of the newly-designed protocols under resource-limited settings. However, the best accuracy is not the target in FL-Block. The original federated learning without deadline limitations achieved accuracies of 0.80 and 0.92 for CIFAR-10 and Fashion-MNIST, respectively. The obtained accuracies are comparable to the performances of FL-Block. Two traditional major concerns, which are the uncertainty of throughput and computation capabilities, do not have a significant impact on the FL-Block's performances.

From the aspect of the impact of T_{round}, we evaluate the changes in the classification accuracy and ToA. As shown in Fig. 3.5. FL-Block and Table 3.3, FL-Block on Fashion MNIST for different values of deadline T_{round} while maintaining T_{final} constant. It is observable that the value of T_{round} should be in a proper range without

Table 3.2 Evaluation results of CIFAR-10

Indexes	ToA(0.5)	ToA(0.75)	Accuracy
T_{round} = 3 sec (r = 0%)	24.5	122.3	0.76
T_{round} = 3 sec (r = 10%)	26.8	136.1	0.74
T_{round} = 3 sec (r = 20%)	30.2	182.5	0.72
T_{round} = 1 sec (r = 0%)	NAN	NAN	0.50
T_{round} = 5 sec (r = 0%)	45.1	178.5	0.77
T_{round} = 10 sec (r = 0%)	80.7	312.5	0.75

ToA(x) (in seconds) is the time costed to reach a specific testing classification accuracy of x. The performance upgrades with the decrease of ToA
Accuracy: denotes the accuracy value for the final iteration

Table 3.3 Evaluation results of fashion-MINIST

Indexes	ToA(0.5)	ToA(0.75)	Accuracy
T_{round} = 3 sec (r = 0%)	15.6	39.3	0.89
T_{round} = 3 sec (r = 10%)	16.6	40.2	0.88
T_{round} = 3 sec (r = 20%)	17.3	45.5	0.91
T_{round} = 1 sec (r = 0%)	5.4	86.4	0.87
T_{round} = 5 sec (r = 0%)	25.1	60.7	0.92
T_{round} = 10 sec (r = 0%)	58.2	110.8	0.93

ToA(x) (in seconds) is the time costed to reach a specific testing classification accuracy of x. The performance upgrades with the decrease of ToA
Accuracy: denotes the accuracy value for the final iteration

being too large or too small. Because of the smaller number of aggregation phases, longer deadlines like 20 s with FL-Block involved numerous end devices in each round, which leads to extremely limited performances. From the other angle, if we set a short deadline like 1 s to limit the number of end devices accessible in each round, the classification performances are also degraded in an unexpected way. Therefore, a possible promising method of selecting T_{round} would be dynamically changing it to align an abundant amount of end devices in each iteration.

3.6.3 Evaluation on Convergence and Efficiency

From the perspective of convergence and efficiency, FL-Block can achieve convergence at different learning rates. Intuitively, greater learning rate results in faster convergence, which is also testified with the evaluation results. As illustrated in Fig. 3.6, FL-Block reaches convergences when data size is moderate, which means the scalability is excellent. Even if the data size is great, the model can still achieve fast convergence. The convergence value of FL-Block is 0.69 when the learning rate is 0.6. It is significant superior comparing to the other two ones, which converges at 0.51 and 0.41, respectively. Therefore, FL-Block fully meets the requirements of big data scenarios, in particular fog computing applications.

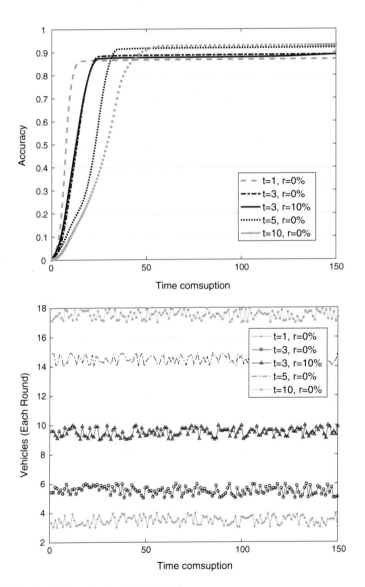

Fig. 3.5 Effects of different values of deadline T_{round}

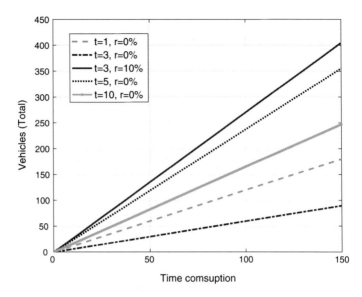

Fig. 3.5 (continued)

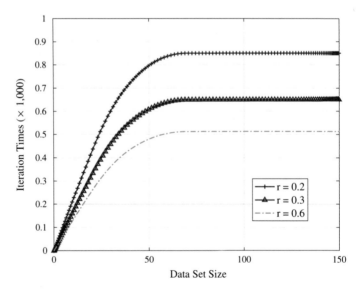

Fig. 3.6 Iteration convergence versus increase of data size

3.6.4 *Evaluation on Blockchain*

In Fig. 3.7, we show how the block generation rate λ influences the average learning completion latency of FL-Block. We can tell that the latency is convex-shaped over the generation rate λ and simultaneously decreasing with the SNRs in Fig. 3.7a. From Fig. 3.7b, we show the minimized average learning completion latency time based

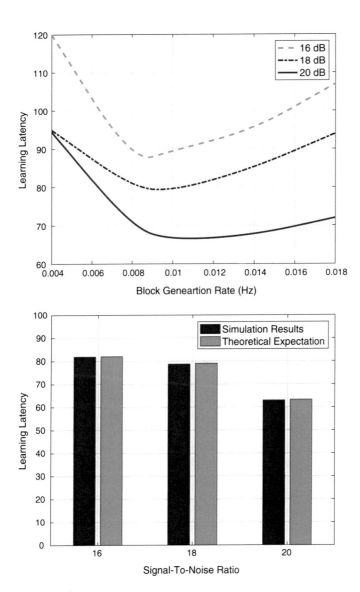

Fig. 3.7 Block generation

on the optimal block generation rate λ^*. The value of latency time derived from the proposition is greater than the simulated minimum latency by up to 2.0%.

Figure 3.8 illustrates the FL-Block's scalability in terms of the numbers N_M and N_V of miners and end devices, respectively. In Fig. 3.8a, the average learning completion latency is computed for $N_M = 5$ and $N_M = 10$ with and without the malfunction of miners. To capture the malfunction, Gaussian noise complying with $N \sim (-0.1; 0.1)$ is added to the aggregated local model updates of each miner with a probability of 0.5. In the scenario of without malfunction, the latency increases with the increase of N_M because of the resource consumption in block propagation delays and cross-verification operation. However, this is not always the case when there is malfunction involved. In FL-Block, the malfunction of each miner only impacts on the global model update of the corresponding end device. This kind of distortion could be eliminated by federating with other non-tempering end devices and miners. On this account, the increase of N_M may result in a shorter latency, which is testified in the provided evaluation results, for instance, $N_M = 20$ with malfunction.

With Fig. 3.8a, we show that a latency-optimal number N_V of end devices are identified. A greater value of N_V enables the usage of a larger number of data samples. However, this will increase the size of each block as well as the delays of block exchange, which brings about the aforementioned convex-shaped latency. It increases each block size such as communication payload. Therefore, it leads to higher block exchange delays and consequently resulting in the convex-shaped latency concerning N_V. That is why a reasonable end device selection can potentially reduce the expected latency.

Lastly, Fig. 3.8b shows that the latency increases with local model size δ_m of each end device, which is intuitive but vital. For this reason, it calls for novel model compression mechanisms, which will also be part of our future work.

3.6.5 Evaluation on Poisoning Attack Resistance

In FL-Block, the poisoning attack could be resisted in a high-confidence way because of the nature of blockchain and the enhanced protocols we propose. In this simulation, we compare the performance of FL-Block with both classic blockchain and federated learning. The assumption here is the initialization phase of the proposed blockchain network where blocks and miners are small enough to allow these attacks to breakthrough Blockchain and thereby poisoning attack is possible to manipulate the data in a malicious way.

In Fig. 3.9a, we use semi-logarithmic coordinate system to compare the performances of the three models. We can tell that with the increase of the adversary's hash-rate, more turbulence there will be for all three models. However, only when the hash-rate reaches a threshold, the adversary can launch a poisoning attack to the data under the protection of Blockchain. This only happens when the number of block is 10 in this simulation. If there are more blocks, it takes much more time as the time consumption increases in an exponential manner.

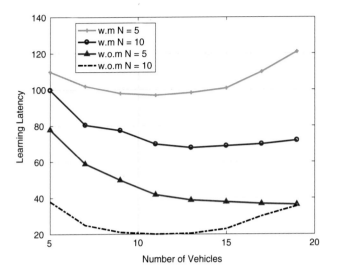

(a) w.m: With Malfunction; w.o.m: Without Malfunction

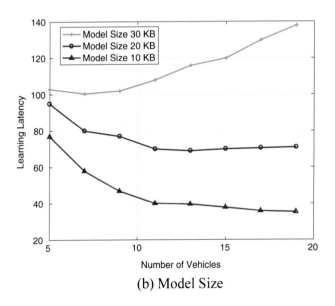

(b) Model Size

Fig. 3.8 Learning latency

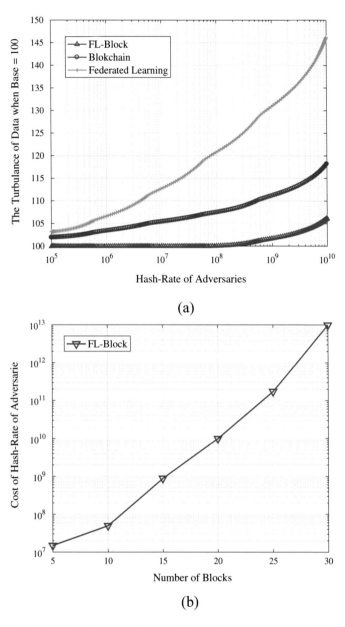

Fig. 3.9 Poisoning attack resistance performance **a** Comparizon among three models, **b** Successful attack requirements based on number of blocks

Figure 3.9b shows how much an adversary needs to breakthrough the protection with regards to different blocks. It looks like in a linearly tendency as the y axis is presented by a semi-logarithmic axis. That means the hash-rate increases in an exponential manner as well. It requires $10^{13}+$ when the block reaches 30. As the difficulty of generating a block increases over time, it provides higher and higher protection when a blockchain develops.

3.7 Summary and Future Work

In this chapter, we propose a novel blockchain-enabled federated learning model to solve the identified issues in IoTs. IoT devices upload the local updates to the edge servers, where the global updates will be generated and stored. Since only the pointer of the global updates is saved on-chain while a distributed hash table (DHT) is used to save to data, the block generation efficiency could be guaranteed. With a hybrid identity generation, comprehensive verification, access control, and off-chain data storage and retrieve, FL-Block enables decentralized privacy protection while preventing single point failure. Moreover, the poisoning attack could be eliminated from the aspect of edge servers. Extensive evaluation results on real-world datasets are presented to show the superiority of FL-Block in IoT scenarios.

For future work, we plan to further extend this model to more generalized scenarios by optimizing the trade-off between privacy protection and efficiency. In addition, we prepare to use game theory and Markov decision process to identify the optimal conditions in terms of computation and communication costs.

References

1. Y. Liu, Y. Qu, C. Xu, Z. Hao, B. Gu, Blockchain-enabled asynchronous federated learning in edge computing. Sensors **21**(10), 3335 (2021)
2. C. Xu, Y. Qu, P.W. Eklund, Y. Xiang, L. Gao, Bafl: an efficient blockchain-based asynchronous federated learning framework, in *2021 IEEE Symposium on Computers and Communications (ISCC)*. IEEE, pp. 1–6 (2021)
3. Y. Qu, L. Gao, T.H. Luan, Y. Xiang, S. Yu, B. Li, G. Zheng, Decentralized privacy using blockchain-enabled federated learning in fog computing. IEEE Int. Things J. (2020)
4. L. Cui, Y. Qu, G. Xie, D. Zeng, R. Li, S. Shen, S. Yu, Security and privacy-enhanced federated learning for anomaly detection in iot infrastructures. IEEE Trans. Indus. Inform. (2021)
5. Y. Wan, Y. Qu, L. Gao, Y. Xiang, Differentially privacy-preserving federated learning using wasserstein generative adversarial network, in *IEEE Symposium on Computers and Communications (ISCC)*. IEEE, pp. 1–6 (2021)
6. C. Mouradian, D. Naboulsi, S. Yangui, R.H. Glitho, M.J. Morrow, P.A. Polakos, A comprehensive survey on fog computing: state-of-the-art and research challenges. IEEE Commun. Surv. Tutor. **20**(1), 416–464 (2018)
7. S. Samarakoon, M. Bennis, W. Saad, M. Debbah, Distributed federated learning for ultra-reliable low-latency vehicular communications, CoRR, arXiv:1807.08127 (2018)

8. Y. Guan, J. Shao, G. Wei, M. Xie, Data security and privacy in fog computing. IEEE Netw. **32**(5), 106–111 (2018)
9. L. Ma, X. Liu, Q. Pei, Y. Xiang, Privacy-preserving reputation management for edge computing enhanced mobile crowdsensing. IEEE Trans. Serv. Comput. **12**(5), 786–799 (2018)
10. M.A. Ferrag, L.A. Maglaras, A. Ahmim, Privacy-preserving schemes for ad hoc social networks: a survey. IEEE Commun. Surv. Tutor. **19**(4), 3015–3045 (2017)
11. Y. Qu, S. Yu, L. Gao, W. Zhou, S. Peng, A hybrid privacy protection scheme in cyber-physical social networks. IEEE Trans. Comput. Soc. Syst. **5**(3), 773–784 (2018)
12. Y. Qu, S. Yu, W. Zhou, S. Peng, G. Wang, K. Xiao, Privacy of things: emerging challenges and opportunities in wireless internet of things. IEEE Wireless Commun. **25**(6), 91–97 (2018)
13. H. Zhu, F. Wang, R. Lu, F. Liu, G. Fu, H. Li, Efficient and privacy-preserving proximity detection schemes for social applications. IEEE Int. Things J. (2017)
14. M.M.E.A. Mahmoud, N. Saputro, P. Akula, K. Akkaya, Privacy-preserving power injection over a hybrid AMI/LTE smart grid network. IEEE Int. Things J. **4**(4), 870–880 (2017)
15. J. Hu, L. Yang, L. Hanzo, Energy-efficient cross-layer design of wireless mesh networks for content sharing in online social networks. IEEE Trans. Vehic. Technol. **66**(9), 8495–8509 (2017)
16. C. Dwork, Differential privacy, in *in Proceedings of ICALP 2006, Venice, Italy, July 10-14, 2006, Proceedings, Part II*, pp. 1–12 (2006)
17. L. Lyu, K. Nandakumar, B.I.P. Rubinstein, J. Jin, J. Bedo, M. Palaniswami, PPFA: privacy preserving fog-enabled aggregation in smart grid. IEEE Trans. Indus. Inform. **14**(8), 3733–3744 (2018)
18. R. Lu, X. Lin, H. Zhu, P. Ho, X. Shen, ECPP: efficient conditional privacy preservation protocol for secure vehicular communications, in *INFOCOM 2008. 27th IEEE International Conference on Computer Communications, Joint Conference of the IEEE Computer and Communications Societies, 13–18 April 2008, Phoenix, AZ, USA*, pp. 1229–1237 (2008)
19. B. McMahan, D. Ramage, Federated learning: Collaborative machine learning without centralized training data. Google Res. Blog **3** (2017)
20. H. Kim, J. Park, M. Bennis, S. Kim, On-device federated learning via blockchain and its latency analysis. CoRR, vol. arXiv:1808.03949 (2018)
21. J. Ni, K. Zhang, Y. Yu, X. Lin, X. S. Shen, Providing task allocation and secure deduplication for mobile crowdsensing via fog computing. IEEE Trans. Dependable Secure Comput. (2018)
22. P. Hu, H. Ning, T. Qiu, H. Song, Y. Wang, X. Yao, Security and privacy preservation scheme of face identification and resolution framework using fog computing in internet of things. IEEE Int. Things J. **4**(5), 1143–1155 (2017)
23. D.X. Song, D.A. Wagner, A. Perrig, Practical techniques for searches on encrypted data, in *2000 IEEE Symposium on Security and Privacy, Berkeley, California, USA, May 14–17, 2000*, pp. 44–55 (2000)
24. R. Lu, X. Lin, T.H. Luan, X. Liang, X.S. Shen, Pseudonym changing at social spots: an effective strategy for location privacy in vanets. IEEE Transa. Vehicu. Technol. **61**(1), 86–96 (2012)
25. H. Li, D. Liu, Y. Dai, T.H. Luan, S. Yu, Personalized search over encrypted data with efficient and secure updates in mobile clouds. IEEE Trans. Emerg. Topics Comput. **6**(1), 97–109 (2018)
26. J. Kang, R. Yu, X. Huang, Y. Zhang, Privacy-preserved pseudonym scheme for fog computing supported internet of vehicles. IEEE Trans. Intell. Transp. Syst. **19**(8), 2627–2637 (2018)
27. H. Tan, D. Choi, P. Kim, S.B. Pan, I. Chung, Secure certificateless authentication and road message dissemination protocol in vanets. Wirel. Commun. Mobile Comput. pp. 7 978 027:1–7 978 027:13 (2018)
28. M. Wazid, A.K. Das, V. Odelu, N. Kumar, M. Conti, M. Jo, Design of secure user authenticated key management protocol for generic iot networks. IEEE Int. Things J. **5**(1), 269–282 (2018)
29. B.S. Gu, L. Gao, X. Wang, Y. Qu, J. Jin, S. Yu, Privacy on the edge: customizable privacy-preserving context sharing in hierarchical edge computing. IEEE Trans. Netw. Sci. Eng. (2019)
30. H. Li, R. Lu, J.V. Misic, M.M.E.A. Mahmoud, Security and privacy of connected vehicular cloud computing. IEEE Netw. **32**(3), 4–6 (2018)

31. L. Ma, Q. Pei, Y. Qu, K. Fan, X. Lai, Decentralized privacy-preserving reputation management for mobile crowdsensing, in *International Conference on Security and Privacy in Communication Systems*. Springer, pp. 532–548 (2019)
32. Y. Jiao, P. Wang, D. Niyato, K. Suankaewmanee, Auction mechanisms in cloud/fog computing resource allocation for public blockchain networks. IEEE Trans. Parallel Distrib. Syst. **30**(9), 1975–1989 (2019)
33. S. Rowan, M. Clear, M. Gerla, M. Huggard, C.M. Goldrick, Securing vehicle to vehicle communications using blockchain through visible light and acoustic side-channels, CoRR arXiv:1704.02553 (2017)
34. N. Ruan, M. Li, J. Li, A novel broadcast authentication protocol for internet of vehicles. Peer-to-Peer Netw. Appl. **10**(6), 1331–1343 (2017)
35. Z. Lu, W. Liu, Q. Wang, G. Qu, Z. Liu, A privacy-preserving trust model based on blockchain for vanets. IEEE Access **6**, 45 655–45 664 (2018)
36. N. Malik, P. Nanda, A. Arora, X. He, D. Puthal, Blockchain based secured identity authentication and expeditious revocation framework for vehicular networks, in *17th IEEE International Conference On Trust, Security And Privacy In Computing And Communications / 12th IEEE International Conference On Big Data Science And Engineering, TrustCom/BigDataSE 2018, New York, NY, USA, August 1–3, 2018*, pp. 674–679 (2018)
37. M. Bernardini, D. Pennino, M. Pizzonia, Blockchains meet distributed hash tables: decoupling validation from state storage, arXiv preprint arXiv:1904.01935 (2019)
38. X. Wang, Y. Han, C. Wang, Q. Zhao, X. Chen, M. Chen, In-edge ai: intelligentizing mobile edge computing, caching and communication by federated learning. IEEE Netw. (2019)
39. V. Smith, C. Chiang, M. Sanjabi, A.S. Talwalkar, Federated multi-task learning, in *Advances in Neural Information Processing Systems 30: Annual Conference on Neural Information Processing Systems 2017, 4–9 December 2017, Long Beach, CA, USA*, pp. 4424–4434 (2017)
40. J. Konecný, H.B. McMahan, D. Ramage, P. Richtárik, Federated optimization: Distributed machine learning for on-device intelligence, CoRR arXiv:1610.02527 (2016)
41. J. Konecný, H.B. McMahan, F.X. Yu, P. Richtárik, A.T. Suresh, D. Bacon, Federated learning: Strategies for improving communication efficiency, CoRR arXiv:1610.05492 (2016)
42. T. Nishio, R. Yonetani, Client selection for federated learning with heterogeneous resources in mobile edge, in *2019 IEEE International Conference on Communications, ICC 2019, Shanghai, China, May 20–24, 2019*, pp. 1–7 (2019)
43. Y. Lu, X. Huang, Y. Dai, S. Maharjan, Y. Zhang, Blockchain and federated learning for privacy-preserved data sharing in industrial iot, IEEE Trans. Indus. Inform. (2019)
44. X. An, X. Zhou, X. Lü, F. Lin, L. Yang, Sample selected extreme learning machine based intrusion detection in fog computing and MEC. Wirel. Commun. Mobile Comput. (2018)
45. Y. Qu, M.R. Nosouhi, L. Cui, S. Yu, Privacy preservation in smart cities, in *Smart Cities Cybersecurity and Privacy*. Elsevier, pp. 75–88 (2019)
46. Y. Qu, S. Yu, W. Zhou, S. Peng, G. Wang, K. Xiao, Privacy of things: emerging challenges and opportunities in wireless internet of things. IEEE Wirel. Commun. **25**(6), 91–97 (2018)
47. J. Yu, K. Wang, D. Zeng, C. Zhu, S. Guo, Privacy-preserving data aggregation computing in cyber-physical social systems. ACM Trans. Cyber-Phys. Syst. **3**(1), 1–23 (2018)
48. L. Cui, G. Xie, Y. Qu, L. Gao, Y. Yang, Security and privacy in smart cities: challenges and opportunities. IEEE Access **6**, 46 134–46 145 (2018)
49. L. Cui, Y. Qu, L. Gao, G. Xie, S. Yu, Detecting false data attacks using machine learning techniques in smart grid: a survey. J. Netw. Comput. Appl., 102808 (2020)
50. L. Gao, T.H. Luan, B. Gu, Y. Qu, Y. Xiang, *Privacy-Preserving in Edge Computing, ser* (Springer, Wireless Networks, 2021)
51. L. Gao, T.H. Luan, B. Gu, Y. Qu, Y. Xiang, Blockchain based decentralized privacy preserving in edge computing, in *Privacy-Preserving in Edge Computing*. Springer, pp. 83–109 (2021)
52. L. Gao, T. H. Luan, B. Gu, Y. Qu, Y. Xiang, Context-aware privacy preserving in edge computing, in *Privacy-Preserving in Edge Computing*. Springer, pp. 35–63 (2021)
53. L. Gao, T.H. Luan, B. Gu, Y. Qu, Y. Xiang, An introduction to edge computing, in *Privacy-Preserving in Edge Computing*. Springer, pp. 1–14 (2021)

54. L. Gao, T.H. Luan, B. Gu, Y. Qu, Y. Xiang, Privacy issues in edge computing, in *Privacy-Preserving in Edge Computing*. Springer, pp. 15–34 (2021)
55. Y. Qu, L. Gao, Y. Xiang, Blockchain-driven privacy-preserving machine learning, *Blockchains for Network Security: Principles, Technologies and Applications*, pp. 189–200 (2020)
56. Y. Qu, M.R. Nosouhi, L. Cui, S. Yu, Personalized privacy protection in big data
57. Y. Qu, M.R. Nosouhi, L. Cui, S. Yu, Existing privacy protection solutions, in *Personalized Privacy Protection in Big Data*. Springer, pp. 5–13 (2021)
58. Y. Qu, M.R. Nosouhi, L. Cui, S. Yu, Future research directions, in *Personalized Privacy Protection in Big Data*. Springer, pp. 131–136 (2021)
59. Y. Qu, M.R. Nosouhi, L. Cui, S. Yu, Leading attacks in privacy protection domain, in *Personalized Privacy Protection in Big Data*. Springer, pp. 15–21 (2021)
60. Y. Qu, M.R. Nosouhi, L. Cui, S. Yu, Personalized privacy protection solutions, in *Personalized Privacy Protection in Big Data*. Springer, pp. 23–130 (2021)
61. M. Series, Guidelines for evaluation of radio interface technologies for imt-advanced, *Report ITU*, vol. 638 (2009)
62. M.R. Akdeniz, Y. Liu, M.K. Samimi, S. Sun, S. Rangan, T.S. Rappaport, E. Erkip, Millimeter wave channel modeling and cellular capacity evaluation. IEEE J. Selected Areas Commun. **32**(6), 1164–1179 (2014)
63. C. Xu, Y. Qu, Y. Xiang, L. Gao, Asynchronous federated learning on heterogeneous devices: a survey, arXiv preprint arXiv:2109.04269 (2021)

Chapter 4
Personalized Privacy Protection of IoTs Using GAN-Enhanced Differential Privacy

The cyber-physical social system (CPSS), as an extension of Internet of Things (IoT), maps human social interaction from cyberspace to the physical world by data sharing and posting, such as publishing spatial-temporal data, images, videos, etc. The published data contains individual's sensitive information, and thereby leads to continues attacks on it. Nowadays, most existing research assumes that the privacy protection should be uniform that all parties share a same privacy protection level. This impractical assumption results in data degradation due to over-protection or privacy leakage due to under-protection. Therefore, we carry out a personalized privacy protection model by using generative adversarial nets (GAN) to enhance differential privacy (P-GAN), especially the protection on sensitive spatial-temporal data. In GAN-DP, we add a *Differential Privacy Identifier* to the classic GAN, in which there are only a *Generator* and a *Discriminator*. P-GAN is able to generate synthetic data can can best mimic the raw data while ensuring a satisfying privacy protection level. It can be told that P-GAN can optimize the trade-off between personalized privacy protection and improved data utility. At the same time, the learning process can achieve fast convergence. Moreover, the GAN-driven differentially private noise generation process decouples the correlation of the injected noise. In this way, collusion attacks, which is a leading attack in privacy-preserving domain, can be eliminated. We evaluate with experiments on real-world datasets, which show the upgradation of optimized trade-off and improved efficiency of P-GAN over existing research. The theories and experiments testify the effectiveness of P-GAN in IoT scenarios. This chapter is mainly based on our recent research on GAN-enhanced differential privacy [1–7].

4.1 Overview

The fast proliferation of the Internet of things (IoT) and increasing demands of real-time social interaction prompt the development of the cyber-physical social system (CPSS). CPSS enables the human interaction from cyberspace to the physical world, which attracts considerable interest from academia, industry, and the public [8]. As a new-emerging extension of social networks, the engagement of users becomes higher progressively while an increasing volume of data is shared among the users for cyber-physical social demands [9].

However, new privacy issues are emerging due to the improper release of sensitive data, in particular, the spatial-temporal data. In a CPSS, the shared spatial-temporal data not only reveals location privacy [10] or trajectory privacy [11] but also potentially leaks the privacy of interest, identity, or even more [12]. Adversaries are incentive to launch continues attacks and benefit from the data with financial interest [13].

Most existing works adopt the method that the privacy protection level is fixed regardless of the real-world requirements or practice demands. For example, dummy-based methods protect privacy against all other users in a universal way [14]. This is not practical because users could be categorized with various indexes such as intimacy. Differential privacy and its current extensions also have the same issue while differential privacy may also lead to insufficient data utility when the privacy requirements are tight [15].

It is essential to improve data utility so as to lift user experience and quality of service in CPSSs [16]. Otherwise, the users will not stick to the services, which will eventually bring the service providers to the brink of collapse. Generative adversarial networks (GAN), which is a practical and popular method to generate synthetic data with some degree of privacy protection, has the potential to leak more sensitive information than the expectation due to the unpredictable randomness [17]. To tackle this, personalization is one potential way to provide the flexibility to privacy protection, which requires less privacy resource cost while slightly improving data utility [18]. However, this is barely discussed in CPSS scenarios.

Despite the significance of personalized privacy, the negative impact is that it is susceptible to collusion attacks, which is a leading attack when multiple parties are involved [19]. Since the users in a specific CPSS receive different variants of data regarding their corresponding intimacy, there is a possibility that different users collude with each other to further breach the privacy [20, 21]. This puts personalized privacy protection under great threats.

All aforementioned issues concurrently pose further challenges to optimize the trade-off between privacy protection and data utility. In order to address these issues, a personalized privacy protection model is proposed by using generative adversarial nets to enhance spatial-temporal private data sharing. The users are categorized by communities and the intimacy is measured by the density of community edge. Then, a QoS-based function will map the intimacy to the privacy protection level. To improve the performances of differential privacy, a modified GAN model is used by adding one more perceptron, namely, *Differential Privacy Identifier*. The modified GAN model has two games running simultaneously, which are between the *Generator*, and the *Discriminator* and *Differential Privacy Identifier*, respectively. In this way, the generated data can perfectly approximate the raw data while satisfying differential privacy requirements. The differentially private data corresponding to intimacy-based the privacy protection level is then shared to other parties with the optimal trade-off between privacy protection and data utility. The convergence of the modified GAN model is also discussed and testified by experimental results.

The contributions of this work are summarized as follows.

- **Personalized Privacy Protection**: A differentially private GAN (P-GAN) model in CPSSs by jointly using differential privacy and GAN. The protection level is personalized by intimacy, which is denoted by the density of community edges in this work. P-GAN complies with ϵ-differential privacy requirements.
- **Optimized Trade-off with Proof against Attacks**: P-GAN achieves an optimized trade-off between personalized privacy protection and improved data utility comparing to current leading methods including dummy-based methods, classic differential privacy, and GAN. Furthermore, the properly decoupled noises eliminate current leading attacks such as the collusion attack in CPSSs.
- **Evaluation on Real-World Data Set**: Extensive experiments on real-world datasets are conducted, and the evaluation results show the performance upgradation from both perspectives of privacy protection and data utility while maintaining efficient convergence.

The rest of this chapter is organized as follows. Section 4.2 presents current research on privacy protection in cyber-physical social systems. The modelling of personalized privacy protection model and collusion attack model are presented in Sects. 4.3 and 4.4, respectively. Section 4.5 shows the security analysis and convergence with optimized data utility, which is followed by evaluations on real-world data sets in Sect. 4.6. At last, the conclusion and future work of this chapter are in Sect. 4.7.

4.2 Related Work

There are mainly four mainstream branches of privacy protection methods, including clustering and dummy, cryptography, machine learning, and differential privacy. In term of clustering-based methods, K-anonymity [22], L-diversity [23], and T-closeness [24] cluster data correspondingly considering magnitude, diversity, and distribution while. But they are not suitable for heterogeneous datasets [25]. Cryptography methods work well, but the computational complexity is relatively high [26, 27]. Besides, it is unfeasible to implement binary access control (trusted and untrusted). The methods based on game theory provide a promising way to optimize the trade-off between privacy and data utility, but the time complexity increase exponentially when multiple parties involve [28]. Differential privacy [15] provides solid theoretical foundation for privacy protection, which have been applied into various applications like vehicular social networks [29, 30].

In this particular scenario, plenty of research has been conducted to preserve the privacy of CPSS. Privacy issues in CPSSs have risen public concerns [31]. In [32], Zhang et al. formulated the trade-off between privacy protection and system performance, in which privacy protection is guaranteed by differential privacy. Qu et al. proposed a decentalized privacy protection method by jointly using blockchain and federated learning. [33]. Yu et al. proposed a privacy-preserving data aggregation model in CPSSs by using bitwise XOR technique to obtain more accurate results [34]. Um et al. proposed a novel method to enhance the trust in CPSSs [35]. Qu et al. devised a novel method that can protect location and identity privacy using the Markov decision process and the Nash equilibrium derived from it [25].

There are some existing works focusing on differentially private machine learning. The difference between this topic and the proposed model is that the proposed one leverages the generative adversarial networks to enhance differential privacy in big data scenario, rather than add privacy protection mechanisms during the machine learning process. Generative adversarial net is proposed by Goodfellow [36]. It requires there are a generator and a discriminator gaming with each other to optimize both perceptrons. Then, Ajovisky et al. [17] proposed Wasserstein generative adversarial networks by extending KL divergence to Wasserstein-1 distance. In [37], the authors devise a triple-GAN by adding an extra perceptron, namely, identifier. Qu et al. and Ho et al. proposed GAN-enabled differential privacy in static and continuous data publishing scenarios, respectively, in [1, 3, 4]. One of the milestones in this field would be [38]. This paper is the early work to use differentially private updates during stochastic gradient descent process. Dwork et al. [39] proposed a novel method to re-use the holdout dataset by differentially private learning. Abadi et al. [40] and Shokri et al. [41] proposed ideas of achieving privacy-preserving deep learning in CCS, respectively. In [2], Acs et al. proposed a differentially private mixture of generative neural networks, which use existing a mixture of generative models to release differentially private learning.

Interested readers are encouraged to explore literature review we have done in recent years [42–46]. Besides, we also summarize several of our book collections to show the frontier of privacy preservation research in edge computing [47–52] and privacy preservation research in edge computing in big data [53–57].

The most similar existing research is conducted by Jordon et al. [58], which proposed a generator with a differential privacy guarantee from a generalized aspect. Different from this work, an extra perceptron is added in the proposed model, namely, differential privacy identifier. The proposed method can improve the accuracy of the generated spatial-temporal data, especially in the cases of cyber-physical social systems.

4.3   Generative Adversarial Nets Driven Personalized Differential Privacy

In this section, the personalized privacy protection model is presented regarding spatial-temporal data published in the cyber-physical social system by means of a modified generative adversarial networks (GAN), which is shown in Fig. 4.1. There are two communities with different community densities. The upper community with a high density observes more accurate data compared with the lower community with a lower density. The rationale is that personalized privacy protection is based on the intimacy level among a group of people, which is denoted by the community density in this work.

Initially, a cyber-physical social network is modelled as a social graph with nodes, edges, and spatiotemporal data. Then, the map is divided into unit regions to accommodate all events and entities. This part also shows how to use a modified GAN algorithm to achieve differential privacy considering continue spatial-temporal data publishing. In addition, a mapping function is devised based on the quality of service (QoS) to generate personalized privacy protection levels. Lastly, the problem is formulated as a min-max problem to derive the optimized trade-off between personalized privacy protection and data utility.

4.3.1   Extended Social Networks Graph Structure

To better represent a social network, graph theory is used, which captures the characteristics of social networks. A single social network can be regarded as $G = \{u_i, e_i, st_i \mid u \in U, e \in E, st \in ST\}$. In the established social graph G, $u \in U$ represents nodes or users, $e \in E$ denotes the edges or relationships, and $st \in ST$ is the spatial-temporal data.

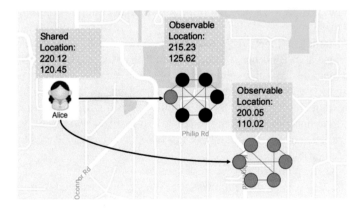

Fig. 4.1 Personalized privacy-preserving data sharing in IoT-based CPSS

For any pair of users (u_i, u_j), if there is at least one path (e.g., $\{e_{i,k_1}, \ldots, e_{k_1,k_n}, e_{k_n,j}\}$) connecting them to each other, there is a relationship between these two users (u_i, u_j). Based on the relationship of the correlated time and space, they are formalized as one integrated index, which is extra sensitive from multiple aspects.

In order to better clarify, a generalized assumption is that the graph G to be an undirected graph. The proposed model still works even if this assumption is removed. Besides, another assumption is the existence of a trusted central authority that processes the data with ϵ-differential privacy and delivers data with privacy-preserving communication. For a trusted central authority, the provided total privacy budget is set to be B and the privacy budget can be assigned to all users based on their corresponding sensitivities in a personalized manner.

In terms of the map division, the definition of the unit region is involved. However, the size of the unit region could be adjusted according to specific real-world scenarios. In the divided map, the event and entity density, the coordinates of the centre, and the estimated radius are required for personalization of privacy protection. The entity density is denoted as the number of entities (restaurants, coffee shops, etc.) divided by the area of the unit region. Similarly, the event density is regarded as the number of all events divided by the number of entities located in this unit region.

4.3.2   GAN with a Differential Privacy Identifier

To protect the continue temporal-spatial data, personalized differential privacy is leveraged in this work. To defeat the primary attacks, a modified GAN is used to generate the noise, with which the correlation among noises could be decoupled. Different from the traditional GAN, one more perceptron is added, namely, differential privacy identifier. This guarantees the generated data both satisfies indistinguishability and complies with differential privacy.

The generator will produce randomized data which is indistinguishable from the raw data. The synthetic data is then compared with raw data by the discriminator. The game process between generator and discriminator supports the improvement of themselves.

The distribution of generator p_g is trained over raw data x_r and a prior of injected noise is defined as $p_z(z)$. Then, the mapping to data space is $G(z; \theta_g)$. $G(z; \theta_g)$ represents the function consisting of multilayer perceptron with θ_g. Another perception, namely, $D(z; \theta_d)$ outputs the probability that x_r is from raw data set rather than p_g. $D(z; \theta_d)$ is trained to maximize the probability of distinguishing the source of the input data. Simultaneously, $G(z; \theta_g)$ is trained to minimize the probability. This confrontation process can also be formulated as a two-player game, and thereby modeled as a Min-Max problem.

During the iteration process, m noise samples $\{z_1, z_2, \ldots, z_m\}$ are selected from $p_g(z)$. In the next step, m data samples $\{x_1, x_2, \ldots, x_n\}$ are selected from $p_d(x)$. The update procedure ascends the discriminator's stochastic gradient with

$$\nabla_{\theta_d} \frac{1}{m} \sum_{i=1}^{m} \left[\log D\left(x_i\right) + \log\left(1 - D\left(G(z_i)\right)\right) \right]. \tag{4.1}$$

In terms of the generator, m noise samples $\{z_1, z_2, \ldots, z_m\}$ from $p_g(z)$ are leveraged to update it with

$$\nabla_{\theta_g} \frac{1}{m} \sum_{i=1}^{m} \log\left(1 - D\left(G(z_i)\right)\right). \tag{4.2}$$

Built upon Eqs. 4.1 and 4.2, the Min-Max problem can be modelled as

$$\min_{G} \max_{D} \mathbb{E}_{x \sim p_d(x)} \left[\log D(x) \right] + \mathbb{E}_{z \sim p_g(z)} \left[\log\left(1 - D\left(G(z_i)\right)\right) \right], \tag{4.3}$$

where $\min_{G}$ minimizes the discrimination probability $\mathbb{P}_g$ while $\max_{D}$ maximizes the successful concealment probability $\mathbb{P}_d$.

$$\begin{cases} \theta_d = \arg\min \left\{ \alpha \cdot L_D\left(t_D, D(x; \theta_D)\right) \right. \\ \qquad\qquad + (1-\alpha) \cdot L_D\left((1-t_D), D(G(z, l; \theta_g); \theta_d)\right) \Big\} \\ \theta_g = \arg\min \left\{ \lambda_t \cdot L_I\left(l, G(z, l; \theta_g)\right) + L_D\left(t_D, D(G(z, l; \theta_g); \theta_d)\right) \right\} \\ \theta_i = \arg\min \left\{ L_I(l, i, \theta_i) \right\} \end{cases} \tag{4.4}$$

To generate differentially private noise with GAN, one more multilayer perceptron is deployed. This perceptron is entitled differential privacy identifier $I(z; \theta_i)$. $I(z; \theta_i)$ is leveraged to identify or classify if synthetic data complying with differential privacy requirements. This identifier perception also confronts with generator $G(z; \theta_g)$ and discriminator $D(z; \theta_d)$. Let be $p_n(n)$ a prior of differentially private data set is defined as $p_n(n)$. The new structure of the interactive discriminator and identifier is represented by $S(D; I)$. The update process can be described as

$$\nabla_{\theta_n} \frac{1}{m} \sum_{i=1}^{m} \left[\log S\left(x_i \big| D; I\right) + \log\left(1 - S\left(G(z_i) \big| D; I\right)\right) \right] \tag{4.5}$$

Extended from conditional generative adversarial net, the proposed model minimizes the cost functions in Eq. 4.4. In Eq. 4.4, l is a binary real-numeral value representing labels of raw data x_r and generator's input data. t_D denotes the label of discriminator with a parameter α.

The proposed model compels differentially private features to be mapped to associated l as an input of the generator. Another parameter λ_t determines a degree that the generator relies on the input features.

The proposed model helps to guide the learning process to generate differential private samples based on the input features rather than only focusing on learning the distribution. Let H be the identification loss of generated samples, λ_t is defined as a learning argument updating with the time step t. λ_t can be described as

$$\lambda_t = \lambda_{t-1} + r \cdot \left\{ L_I\left(l, G(z, l; \theta_g)\right) - E \cdot L_I(l, x) \right\}, \qquad (4.6)$$

where r is defined as learning rate of λ_t.

In the proposed model, there are two confrontations gaming simultaneously. In addition to the game between generator and discriminator, the confrontation between differential privacy identifier and discriminator is another game process. To balance discriminator and identifier, E is utilized to maintain the Nash Equilibrium of both targets. Assume there exists a perfectly-trained generator $G(z, l)$ obtaining an existing true distribution, and then a group of generated noises shares the same identification loss comparing to the raw data set. Therefore, $E = \frac{L_I(l, G(z, l; \theta))}{L_I(L_I(l, x))=1}$ when $G(z, l) = P(x)$. When the generator is trained to take input features into consideration, the value of E would be less than 1.

By adding the new perceptron, the Min-Max problem can be reformulated as

$$\min_{G} \max_{S} \; \mathbb{E}_{x \sim p_d(x)}\left[\log S(x|D; I) \right]$$
$$+ \mathbb{E}_{z \sim p_z(z)}\left[\log\left(1 - S\left(\left(G(z_i)\right) \middle| D; I \right) \right) \right]. \qquad (4.7)$$

Algorithm 1 Generative Adversarial Net Enabled Differential Privacy

---

Input: $d(x)$, r, and ϵ;
Output: Differentially private synthetic data set follows $p_g = Pr(GAN(p_x))$;
1: **while** $p_g \neq Pr(GAN(p_x))$ **do**
2:    **for** l iterations **do**
3:       $\{z_1, z_2, ..., z_m\}$ from $p_g(z)$;
4:       $\{x_1, x_2, ..., x_m\}$ from $d(x)$;
5:       Update the D by ascending its stochastic gradient;
6:       **for** k iterations **do**
7:          $\{x_1, x_2, ..., x_m\}$ from $d(x)$;
8:          Update the I by ascending its stochastic gradient;
9:       **end for**
10:    **end for**
11:    $\{x_1, x_2, ..., x_m\}$ from $d(x)$;
12:    Update G by ascending its stochastic gradient;
13: **end while**

4.3.3 Mapping Function

The personalized privacy protection is based on the level of intimacy. Therefore, a social distance is adopted to denote the level of intimacy. In this work, the social distance is denoted by the density of the community. The density of the community is represented as the number of friend pairs divided by the number of all possible friend pairs.

In this work, friendship is used structure to accomplish personalized privacy instead of traditional methods that leverage social distance. The social distance is hard to measure and sometimes not practical. The reason is that social distance takes irrelative users into consideration, which leads to additional computational cost without improvement of privacy protection.

The friendship is measured with the edge density of community. If there are more users in an identified community, the privacy protection will be released to a proper level, vice versa. The rationale is the connection of user relationships. People tend to make good friends with a group of friends in real-world and then form a community in cyber-physical social networks. Based on this, privacy protection level is personalized according to community parameters. Therefore, the community detection is based on *link communities*.

Algorithm 2 Edge Pairs-based Link Community Detection Algorithm

---

Input: A CPSS;
Output: Detected communities $C = \{P_1, P_2, ..., P_c\}$;
1: Initialize $e_i \in E$ as a community;
2: Initialize amount of neighbours $n_+(i)$;
3: Calculate $S\left(e_{ij}, e_{ik}\right)$ with $n_+(j)$ and $n_+(k)$;
4: Sort $S\left(e_{ij}, e_{ik}\right)$ in descending order;
5: Merge the communities with a tree structure;
6: **while** $D < D_t$ **do**
7:     Calculate M and $|C|$;
8:     Calculate m_c and n_c as edge pair number and node number in P_c;
9:     Derive partitioning density of P_c with m_c and n_c;
10:    Update partitioning density as $D = \frac{1}{M} \sum_c m_c D_c$;
11: **end while**
12: Transform edge pair-based tree graph into node-based graph;
13: Output $C = \{P_1, P_2, ..., P_c\}$ and $u_i \in U$;

A set of closely correlated edges $C = \{e_{i,j} | e \in E\}$ are used to describe a community instead of defining it with a set of strongly interlinked nodes $C = \{u_i | u \in U\}$. The advantage is that an edge normally corresponds to a specific single community $\{e_i \in C_i\}$ while nodes actually can involve in multiple communities $\{u_i \in U_{i_1}, U_{i_2}, ..., U_{i_n}\}$. In this way, the overlap nodes problem can be solved. In addition, through different threshold cut trees, the hierarchy link community structure can be generated.

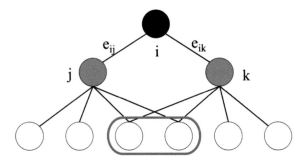

Fig. 4.2 Overlapped nodes and edge similarity: Nodes could be in multiple communities since the edge is used to form the communities

At first, each edge is seen as a community. Then, other edges are enrolled into this community, which shares the same common nodes with the first edge. The similarity of an edge pair (e_{ij}), e_{ik} with a common node u_i is to consider the similarity of u_k and u_j, which is shown in Fig. 4.2. One straightforward way to achieve this is to measure the neighbour's amounts of the u_k and u_j. Therefore, the similarity of (e_{ij}), e_{ik} is formulated as

$$S\left(e_{ij}, e_{ik}\right) = \frac{\left|n_+(k) \cap n_+(j)\right|}{\left|n_+(k) \cup n_+(j)\right|}, \tag{4.8}$$

where $n_+(k)$ is the set of node u_k and all its adjacent neighbours while $n_+(j)$ is the set of node u_j and all its adjacent neighbours

With the similarity of edge pairs, the cyber-physical social network community structures are detected with this hierarchy clustering method. Firstly, the similarities $S(\cdot)$ of all the edge pairs (e_{ij}), e_{ik} are calculated inside the social network and sort the values of similarities in descending order. Secondly, the communities merge based on the ordered edge pairs and represent the merging process with a tree graph structure. In this step, if there are some edge pairs sharing the same similarity, they can be merged at the same time. Thirdly, the operation needs a threshold, or all the edge pairs will merge into a single community eventually.

During the above procedure, the similarity leveraged to combine two communities is also known as the strength of a merged community, which also corresponds to the height of the tree graph's branch. For the purpose of obtaining the best community structure, it is necessary to determine the best position to cut the tree graph, or equivalently, confirm the threshold of the community merging process. Therefore, an objective function is defined, namely, partition density $D(\cdot)$, based on the density of edge pairs inside communities.

Let M be the number of edge pairs inside a cyber-physical social network, $|C|$ be the number of communities $\{P_1, P_2, ..., P_c\}$, m_c and n_c are the number of edge pairs and nodes inside community P_c, the corresponding normalization density is

$$D_c = \frac{m_c - \left(n_c - 1\right)}{n_c\left(n_c - 1\right)/2 - \left(n_c - 1\right)}, \tag{4.9}$$

where $n_c - 1$ is the minimum number of edge pairs required to constitute a connected graph and $n_c(n_c - 1)/2$ is maximum number of possible edge pairs among n_c nodes. A special consideration is that $D_c = 0$ if $n_c = 2$. Thus, the partition density of the whole network is formulated as the weighted sum of D_c.

$$D = \frac{1}{M} \sum_c m_c D_c = \frac{2}{M} \sum_c m_c \frac{m_c - \left(n_c - 1\right)}{\left(n_c - 2\right)\left(n_c - 1\right)} \tag{4.10}$$

As every term in the summation is limited inside the community, the model avoids the distinguishability limitation problem of modularity. The satisfying community detection are obtained through calculating partition density corresponding to each level or optimize the partition density directly. The most primary advantage of edge pair is to derive the best partitioning position with maximum density. Another advantage of using edge pairs tree graph is to reveal the hierarchy community structure feature by non-optimization cut-off rule.

Generally, the obtained edge pair community is reformed back to the node-based community, which is shown in Fig. 4.3. In this way, the overlap nodes can be easily found. Although an edge can only belong to one edge pair community, the multiple implications of the edge can be observed even if its two end nodes belong to two same communities simultaneously.

After obtaining the social distance d_{ij}, a Sigmoid function $S(\cdot)$ is used to map social distance to privacy level ϵ. The sigmoid function is widely used to measure users' satisfaction degrees in terms of quality of service (QoS), which captures the characteristics of social networks. The rationale behind Sigmoid function is that the privacy protection level remains high when the social distance grows in a low range. The privacy protection level decreases quickly when the social distance grows across a certain threshold. The further improvement of privacy level become marginal and produces little benefits when social distance grows in a high range. The mapping function is illustrated as

$$\epsilon_i = k \times \frac{1}{1 + \exp(-\theta \cdot (d_{ij}) - m)}, \tag{4.11}$$

where k is the weight parameter to adjust the amplitude of the maximum value, θ is leveraged to decide the steepness of the curve, and m denotes the location of the

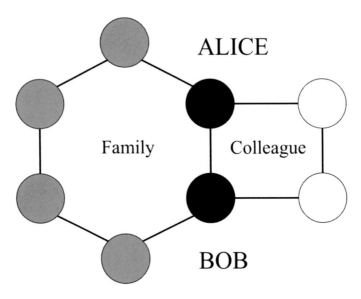

(a) Edge Pair Community

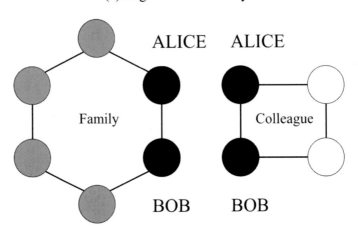

(b) Node Community

Fig. 4.3 Edge pair link community to node community: The most primary advantage of edge pair is to derive the best partitioning position with maximum density. Another advantage of using edge pairs tree graph is to reveal the hierarchy community structure feature by non-optimization cut-off rule

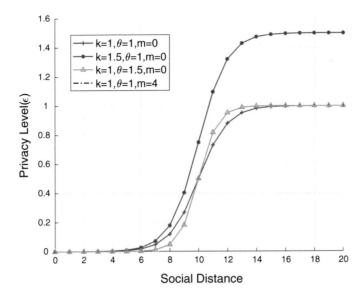

Fig. 4.4 Curve transformation of sigmoid function with different parameter: sigmoid function is widely used to measure users' satisfaction degree in term of quality of service (QoS), which captures the characteristics of CPSS

symmetric line. In Fig. 4.4, the curve transformation shows the change of curves regarding each parameter. The figure testifies that Sigmoid function is appropriate for mapping social distance to privacy protection level from the perspective of QoS.

4.3.4 Opimized Trade-Off Between Personalized Privacy Protection and Optimized Data Utility

In a personalized privacy-preserving data publishing problem, there are different approximations $\{y_{ij}\}_{j=1}^{n}$ resulting in multiple values of data utility. Therefore, when the optimized utility is mentioned in this work, the sum of data utility is referred to as Eq. 4.12 displays.

$$\sum_{i=1}^{n}\sum_{j\neq i}^{n}E\left|\left|y_{ij}-d_i\right|\right|_2^2 \tag{4.12}$$

Given maximum privacy level and minimum data utility $\min(DU)$ and the optimized tradeoff is modelled as Eq. 4.13.

$$\text{Objective}: \min \sum_{i}^{n} \epsilon_i$$

$$s.t.$$

$$\epsilon_i = k \times \frac{1}{1 + \exp(-\theta \cdot (d_{ij}) - m)} \tag{4.13}$$

$$\sum_{i=1, j \neq i}^{n,n} M_{DP}\left(\epsilon\left(\frac{1}{d_{ij}}\right)\right) \leq \max M_{DP}\left(\epsilon\left(\frac{1}{d_{ij}}\right)\right),$$

$$\sum_{i=1}^{n} \sum_{j \neq i}^{n} E\left\|y_{ij} - d_i\right\|_2^2 \geq \min(DU)$$

4.4 Attack Model and Mechanism Analysis

The adversary model with the corresponding attack model are discussed for further analysis. Instead of measuring the adversaries and attacks in a qualitative way, a quantitative model is established within the framework of differential privacy.

In terms of adversaries, an adversary with background knowledge is considered, which is practical and feasible in real-world applications. From the perspective of attacks, the collusion attack is considered, which is one of the leading attacks when multiple parties are involved. In CPSS, the most pivotal background knowledge and internal correlation is identified, namely, spatial-temporal data.

4.4.1 Collusion Attack

Similar to adversaries with background knowledge, the collusion attack suffers from the difficulties of quantitative modeling. Thanks to differential privacy, the collusion attack can also be modeled under its framework.

Traditionally, the collusion attack is described as an attack launched with multiple data resources. Followed by this proposition, the collusion attack is further defined as

Definition (Collusion Attack)
Given multiple data resources $\{D_i | i = 1, 2, 3, ..., n\}$, an collusion attack $L(\cdot)$, and M to be a randomized algorithm that sanitizes the dataset where $\epsilon_i = M(D_i)$, the linkage attack is successfully launched if

$$\sum_{i}^{n} M(D_i) \geq L(\cdot) \geq \max\left\{M(D_i) \middle| i = 1, 2, ..., n\right\} \tag{4.14}$$

The successful collusion attack is regarded as the release of the privacy protection level. Therefore, there is a range for the increased ϵ. In this work, the worst case is consided, followed by the protection algorithm which minimizes the damage of collusion attack.

4.4.2 Attack Mechanism Analysis

In a collusion attack modeled by differential privacy, it can be analyzed by the composition mechanism, which is a key mechanism in the framework of differential privacy.

The composition mechanism provides the functionality that various random mechanisms can work together. With the composition mechanism, more complicated and secure mechanisms can be devised by utilizing the advantages of various mechanisms.

Multiple random mechanisms $\{M_1, M_2, ..., M_n : \mathcal{D} \to \Delta(\mathcal{Y})\}$ respectively satisfies $\{\epsilon_1, \epsilon_2, ..., \epsilon_n\}$-differential privacy. If the n mechanisms are composed together to process the same data, the composed mechanism is $\sum_i^n \epsilon_i$-differentially private.

When personalized privacy is used to process a piece of data, this data is definitely processed with various mechanisms with different privacy levels $\{\epsilon_i | i = 1, 2, 3, ..., n\}$. That is a change for the adversary to obtain data from various sources and release the privacy level by the composition feature.

An example of two-fold personalized privacy protection levels is presented for clarity. Given two privacy levels ϵ_i and ϵ_{i+1}, where $\epsilon_{i+1} > \epsilon_i$. There is a mechanism $M_{\epsilon_i \to \epsilon_{i+1}} : \mathcal{D} \to \Delta(\mathcal{Y}^2)$ which releases the data in two different social networks. At first, the u_i publishes a noisy outcome y_{ij}^1 to u_j. The y_{ij}^1 satisfies ϵ_i-DP. Afterwards, the privacy protection level is relaxed to ϵ_{i+1}-DP. In case that users collude with other to obtain more precise output, the proposed mechanism should at least satisfy

$$M_{DP}\left(\epsilon_i + \epsilon'_{i+1}\right) = M_{DP}\left(\epsilon_{i+1}\right), \tag{4.15}$$

where ϵ'_{i+1} is the privacy level of the second noisy response and M_{DP} denotes the differentially private mechanism. As the upper bound of composition mechanism indicates, Eq. 4.16 is satisfied.

$$M_{DP}\left(\epsilon'_{i+1}\right) = M_{DP}\left(\epsilon_{i+1} - \epsilon_i\right), \tag{4.16}$$

where $\epsilon'_{i+1} < \epsilon_{i+1}$. This conclusion implies the second noisy response can not relax the privacy protection level to a satisfying degree. Especially in the case that $\epsilon(1) < \epsilon_{i+1} \ll 1$, the privacy level may even be upgraded. The data utility degrades so that it is not suitable for practical applications.

4.5 System Analysis

It is vital for a privacy protection system to keep a balance between data utility and the level of privacy protection. To maintain data utility, the system must preserve the accuracy of privacy-aware responses. For this reason, the optimal amount of noise should be injected into the users' private location regarding the trade-off between privacy protection and data utility. In other words, the noise magnitude must not be more than what is required for privacy protection.

In the proposed mechanism, the accuracy of response L_{ij} can be measured by squared error $\triangle_{ij}$ as

$$\triangle_{ij} = ||L_{ij} - L_i||_2^2 \quad i, j \in V \tag{4.17}$$

where a smaller error represents more accuracy. Then it can be shown as

$$
\begin{aligned}
L_{ij}^{(}1) &= L_i^{(}1) + N_1(\epsilon(d_{ij})) \\
L_{ij}^{(}2) &= L_i^{(}2) + N_2(\epsilon(d_{ij}))
\end{aligned}
\tag{4.18}
$$

where $L_{ij}^{(}k) \in R(k = 1, 2)$ are the GPS coordinates of response L_{ij} and $N_k(\epsilon(d_{ij}))(k = 1, 2)$ are independent and identically distributed random variables, i.e.

$$N_k \sim P - GAN(0, \epsilon(d_{ij})) \tag{4.19}$$

Therefore, the squared error $\triangle_{ij}$ is obtained as

$$\triangle_{ij} = N_1^2(\epsilon(d_{ij})) + N_2^2(\epsilon(d_{ij})) \tag{4.20}$$

Because $N_1^2 + N_2^2 = \triangle$ corresponds to a circle with radius $\sqrt{(\triangle)}$, the cumulative distribution function of $\triangle$ is formulated as

$$
\begin{aligned}
F_\triangle(\delta) &= Pr[\triangle \le \delta] = Pr[(N_1^2 + N_2^2) \le \delta] \\
&= \int_{-\sqrt{\delta}}^{\sqrt{\delta}} \int_{-\sqrt{\delta - n_2^2}}^{\sqrt{\delta - n_2^2}} f_{N_1, N_2}(n_1, n_2) dn_1 dn_2
\end{aligned}
\tag{4.21}
$$

Since N_1 and N_2 are independent and identically distributed, then it is shown as

$$f_{N_1, N_2}(n_1, n_2) = f_{N_1}(n_1) f_{N_2}(n_2) = \frac{\epsilon^2}{4} e^{-\epsilon(|n_1 + n_2|)} \tag{4.22}$$

Therefore,

$$F_\Delta(\delta) = \int_{-\sqrt{\delta}}^{\sqrt{\delta}} f_{N_2}(n_2) \int_{-\sqrt{\delta-n_2^2}}^{\sqrt{\delta-n_2^2}} \frac{\epsilon}{2} e^{-\epsilon|n_1|} dn_1$$

$$= \frac{\epsilon}{2} \int_{-\sqrt{\delta}}^{\sqrt{\delta}} (1 - e^{-\epsilon\sqrt{\delta-n_2^2}}) e^{\epsilon|n_2|} dn_2$$

(4.23)

By taking differentiation, the probability density function (PDF) of Δ is

$$f_\Delta(\delta) = \frac{d}{d\delta} F_\Delta(\delta) = \frac{\epsilon}{2\sqrt{\delta}} e^{-\epsilon\sqrt{\delta}}$$

(4.24)

From above equation it is derived that Δ has generalized gamma distribution with scale parameter $\frac{1}{\epsilon^2}$, expected value $\frac{2}{\epsilon^2}$ and variance $\frac{20}{\epsilon^4}$. This means that the random variable Δ depends only on ϵ, i.e. for larger values of ϵ (equivalently, smaller social distances), with high probability, a smaller Δ is offered and vice versa. Therefore, the accuracy of the responses L_{ij} is determined by the privacy level ϵ only and they have a direct relation, i.e. an increase in ϵ results in a more accurate response. This is exactly what the mechanism needs to satisfy both flexible (variable) privacy and optimal data utility.

4.6 Evaluation and Performance

This evaluation section demonstrates the experiments and results derived from real-world datasets. Multiple indexes, including privacy protection level, data utility, and efficiency are demonstrated to evaluate the performances. The proposed model (P-GAN) is compared with a dummy-based method (Dummy), classic Laplace mechanism (Laplace), and Generative Adversarial Nets (GAN). The evaluation results show the superiority of the proposed model and confirm the significance of this work. The algorithms are implemented using Python and are executed on Mac OS platform with Core I5@2.7GHz CPU and 8G memory.

The proposed model is evaluated on the Yelp challenge dataset round 2019 [59] released by Yelp Inc. In this dataset, the information of business, user, and the reviews are used, which correspond to location and identity in our model, respectively. The details of the Yelp dataset are as follows. The data is from metropolitan areas. There are totally $6.68M$ reviews and $1.22M$ tips from $1.63M$ users for $192K$ businesses. In the business part, there are over $1.2M$ business attributes, e.g., hours and parking availability.

The dummy-based method is the most fundamental and feasible method in practice. The key idea of this method is to generate dummies with various techniques. Laplace mechanism is a classic mechanism of differential privacy, with which controllable noise will be injected and a noisy output will be shared with the recipients. Generative adversarial nets is a generative model which generates a similar entity by the game between a generator and a discriminator.

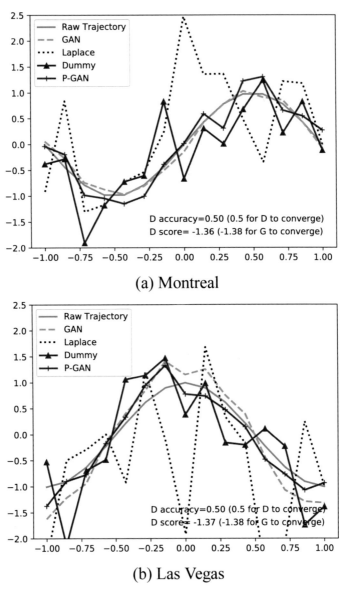

(a) Montreal

(b) Las Vegas

Fig. 4.5 Evaluation on the generated trajectory: In all three figures, the performances of P-GAN against that of GAN, Laplace, and Dummy-based method are demonstrated. It is clear that P-GAN and GAN can imitate the trajectories well while P-GAN can maintain higher level of privacy protection guaranteed by differential privacy

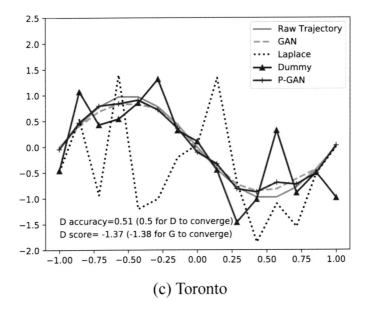

(c) Toronto

Fig. 4.5 (continued)

4.6.1 Trajectory Generation Performance

In Fig. 4.5, the generated trajectories using all mainstream methods are shown based on real trajectories from three different cities, including *Montreal, Las Vegas, and Toronto*. As shown in figures, the dummy-based method and Laplace mechanism of differential privacy can hardly imitate the real-world trajectories. Both GAN and the proposed PGAN can generate excellent approximations while P-GAN can provide higher quantitative privacy protection while maintaining satisfying data utility (see Sects. 4.6.2 and 4.6.3).

4.6.2 Personalized Privacy Protection

In Fig. 4.6, the performance of personalized privacy protection is presented. In Fig. 4.6a, it can be told that the privacy protection levels of the dummy-based method are 0.81 and not changing with the increase of social distance, which is denoted by community density in this work. The privacy protection levels of GAN fluctuates from 0.64 to 0.72 randomly. Although there is some flexibility in GAN, GAN cannot adapt itself to different specific scenarios. In the case of the Laplace mechanism, it shares the same flexible privacy protection with P-GAN in this case. The privacy levels drop from 0.96 to 0.27 with the increase of social distance. However, the data utility of P-GAN is higher as shown in Sect. 4.6.3.

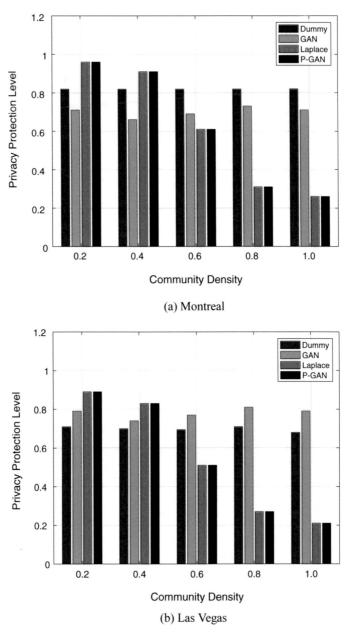

(a) Montreal

(b) Las Vegas

Fig. 4.6 Evaluation on personalized privacy protection level: In term of personalized privacy protection level, P-GAN and Laplace could achieve it in a flexible way. Dummy-method can only provide uniform privacy protection and GAN protect the privacy around a certain level

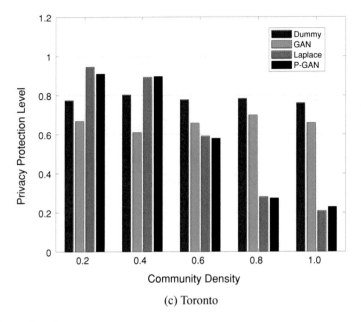

(c) Toronto

Fig. 4.6 (continued)

4.6.3 Data Utility

The evaluation results of data utility are shown in Fig. 4.7. From the line chart, the observation is that the Laplace mechanism is the most unstable one and increases from 0.30 to 0.84 in a log-like manner. The performance of the dummy-based method is steady but the ceiling of data utility is 0.80, which is not sufficient. GAN is not affected by the privacy parameter ϵ and guarantees high data utility but lacks of enough privacy protection as discussed in the previous subsection. P-GAN has the second-best performance in terms of data utility while the performance upgrades from 0.71 to 0.93 with an increase of ϵ, which makes it easily extendable and practical.

4.6.4 Efficiency and Convergence

Figure 4.8 shows the convergence of each method. Since the dummy-based method and Laplace mechanism are highly dependent on the pre-defined parameters without updating process, they maintain a relatively steady status independent of the time or iteration. In Montreal, the convergence indexes of them both fluctuate around 0.8. In terms of GAN and P-GAN, both of them can converge at the learning rate of 0.01 and iteration times of 1300. The cases of Toronto and Las Vegas converges at the

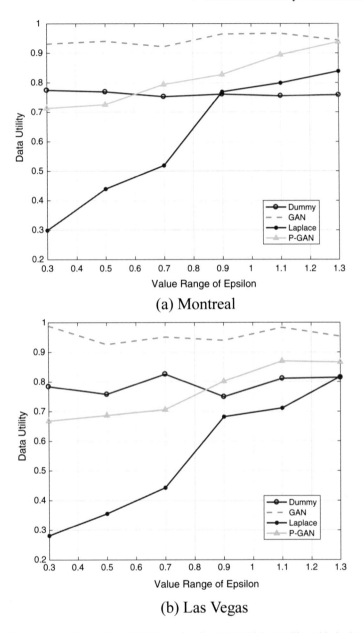

(a) Montreal

(b) Las Vegas

Fig. 4.7 Evaluation on data utility: P-GAN can finally achieve highest utility with the increase of ϵ. GAN maintains high data utility regardless of ϵ but has no significant advantages after $\epsilon > 1.2$. Dummy-based method highly depends on the initial settings while Laplace performs poorly when $\epsilon < 0.9$

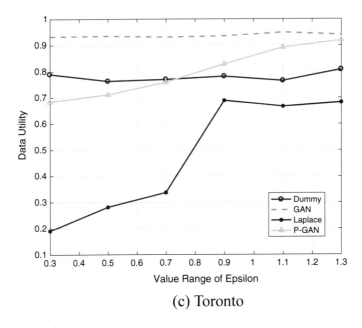

(c) Toronto

Fig. 4.7 (continued)

iteration times of 1200 and 1400, respectively. Therefore, P-GAN does not require extra computational resources comparing to GAN. This feature enables its scalability in this big data era.

4.6.5 Further Discussion

By comprehensively measuring the performances of P-GAN, the conclusion is that P-GAN can integrate the advantages of differential privacy and GAN while eliminating the negative impact of these two models. Comparing to differential privacy, P-GAN can maintain the same privacy protection level while significantly improving the data utility. As to GAN, P-GAN can achieve the same data utility after several thousands of iterations while providing a higher privacy guarantee. Regarding efficiency, P-GAN can maintain efficient processing in the scenario of CPSS as the data sharing frequency is not high. All in all, P-GAN can achieve the optimized trade-off between privacy protection and data utility by taking advantage of both differential privacy and GAN with *negligible* efficiency sacrifice.

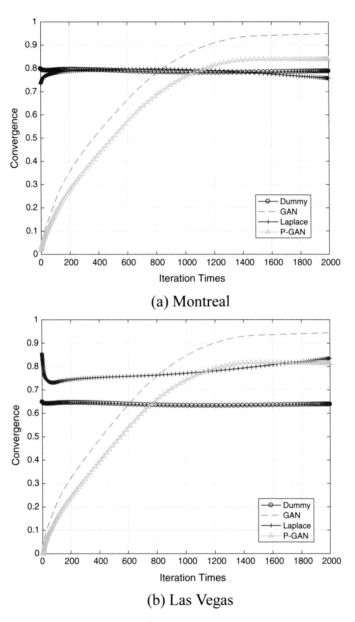

(a) Montreal

(b) Las Vegas

Fig. 4.8 Evaluation on efficiency and convergence: All methods can reach an convergence after certain rounds of iterations. P-GAN converges at a similar rate comparing to GAN, which means the efficiency is satisfying. Laplace and dummy-based method maintains steady due to lack of learning features

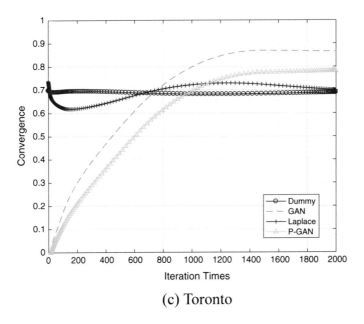

(c) Toronto

Fig. 4.8 (continued)

4.7 Summary and Future Work

In the context of IoT, most existing privacy-preserving research fails to balance privacy protection and spatial-temporal data utility. To address this issue, a personalized model using GAN-enabled differential privacy is proposed to derive the optimized trade-off, especially for spatial-temporal privacy-preserving data sharing. Both theoretical analysis and experimental results on three sub-datasets confirm that the proposed model can achieve an optimized trade-off between privacy protection and data utility compared with current mainstream models while maintaining a satisfying convergence speed. This shows the effectiveness and superiority of the proposed model. This work provides insights to enable differential privacy using GAN or potentially other machine learning paradigms such as federated learning. In addition, this work advances the performances of differential privacy in this particular scenario. The limitation of this model may include the differential privacy is guaranteed by resource-intensive learning algorithm (GAN), which leads to low efficiency and high computing resource consumption. For future work, the first plan is to improve the efficiency of the proposed model by possible reduction of cardinality. In addition, differentially private federated learning is another potentially promising direction to follow in IoT scenarios.

References

1. S. Ho, Y. Qu, L. Gao, J. Li, and Y. Xiang, Generative adversarial nets enhanced continual data release using differential privacy, in *International Conference on Algorithms and Architectures for Parallel Processing*. Springer, pp. 418–426 (2019)
2. G. Ács, L. Melis, C. Castelluccia, E.D. Cristofaro, Differentially private mixture of generative neural networks, in *2017 IEEE International Conference on Data Mining, ICDM 2017, New Orleans, LA, USA, November 18-21, 2017*, pp. 715–720 (2017)
3. Q. Youyang, Z. Jingwen, L. Ruidong, Z. Xiaoning, Z. Xuemeng, Y. Shui, Generative adversarial networks enhanced location privacy in 5g networks. Sci China Inform Sci
4. Y. Qu, S. Yu, J. Zhang, H.T.T. Binh, L. Gao, W. Zhou, Gan-dp: generative adversarial net driven differentially privacy-preserving big data publishing, in *ICC 2019-2019 IEEE International Conference on Communications (ICC)*. IEEE, pp. 1–6 (2019)
5. Y. Qu, S. Yu, W. Zhou, Y. Tian, Gan-driven personalized spatial-temporal private data sharing in cyber-physical social systems. IEEE Transa. Netw. Sci. Eng. **7**(4), 2576–2586 (2020)
6. L. Cui, Y. Qu, G. Xie, D. Zeng, R. Li, S. Shen, S. Yu, Security and privacy-enhanced federated learning for anomaly detection in iot infrastructures. IEEE Trans. Indus. Inform. (2021)
7. Y. Wan, Y. Qu, L. Gao, Y. Xiang, Differentially privacy-preserving federated learning using wasserstein generative adversarial network, in *2021 IEEE Symposium on Computers and Communications (ISCC)*. IEEE, pp. 1–6 (2021)
8. A. Sheth, P. Anantharam, C. Henson, Physical-cyber-social computing: an early 21st century approach. IEEE Intell. Syst. **28**(1), 78–82 (2013)
9. D. Garcia, Leaking privacy and shadow profiles in online social networks. Sci. Adv. **3**(8), e1701172 (2017)
10. A.R. Beresford, F. Stajano, Location privacy in pervasive computing. IEEE Pervasive Comput. **2**(1), 46–55 (2003)
11. C. Chow, M.F. Mokbel, Trajectory privacy in location-based services and data publication, *SIGKDD Explorations*, vol. 13, no. 1, pp. 19–29 (2011). https://doi.org/10.1145/2031331.2031335
12. Y. Qu, S. Yu, W. Zhou, S. Peng, G. Wang, K. Xiao, Privacy of things: emerging challenges and opportunities in wireless internet of things. IEEE Wirel. Commun. **25**(6), 91–97 (2018)
13. S. Yu, Big privacy: challenges and opportunities of privacy study in the age of big data. IEEE Access **4**, 2751–2763 (2016)
14. H. Liu, X. Li, H. Li, J. Ma, X. Ma, Spatiotemporal correlation-aware dummy-based privacy protection scheme for location-based services, in *2017 IEEE Conference on Computer Communications, INFOCOM 2017, Atlanta, GA, USA, May 1–4, 2017*, pp. 1–9 (2017)
15. C. Dwork, Differential privacy, in *in Proceedings of ICALP 2006, Venice, Italy, July 10-14, 2006, Proceedings, Part II*, pp. 1–12 (2006)
16. L.T. Yang, X. Wang, X. Chen, L. Wang, R. Ranjan, X. Chen, M.J. Deen, A multi-order distributed hosvd with its incremental computing for big services in cyber-physical-social systems, IEEE Trans. Big Data (2018)
17. M. Arjovsky, S. Chintala, L. Bottou, Wasserstein generative adversarial networks, in *Proceedings of the 34th International Conference on Machine Learning, ICML 2017, Sydney, NSW, Australia, 6–11 August 2017*, pp. 214–223 (2017)
18. Z. Jorgensen, T. Yu, G. Cormode, Conservative or liberal? personalized differential privacy, in *31st IEEE International Conference on Data Engineering, ICDE 2015, Seoul, South Korea, April 13-17, 2015*, pp. 1023–1034 (2015)
19. Y. Liu, N. Li, Retrieving hidden friends: a collusion privacy attack against online friend search engine. IEEE Trans. Inform. Forensics Secur. **14**(4), 833–847 (2019)
20. X. Zheng, Z. Cai, J. Yu, C. Wang, Y. Li, Follow but no track: privacy preserved profile publishing in cyber-physical social systems. IEEE Int. Things J. **4**(6), 1868–1878 (2017)
21. Q. Xu, P. Ren, H. Song, Q. Du, Security-aware waveforms for enhancing wireless communications privacy in cyber-physical systems via multipath receptions. IEEE Int. Things J. **4**(6), 1924–1933 (2017)

22. S. Pierangela, S. Latanya, Protecting privacy when disclosing information: k-anonymity and its enforcement through generalization and suppression, in *Proceedings of the IEEE Symposium on Research in Security and Privacy*, pp. 1–19 (1998)

23. A. Machanavajjhala, D. Kifer, J. Gehrke, M. Venkitasubramaniam, L-diversity: privacy beyond k-anonymity. IEEE Trans. Knowl. Data Eng. **1**(1) (2007)

24. N. Li, T. Li, S. Venkatasubramanian, Closeness: a new privacy measure for data publishing. IEEE Trans. Knowl. Data Eng. **22**(7), 943–956 (2010)

25. Y. Qu, S. Yu, L. Gao, W. Zhou, S. Peng, A hybrid privacy protection scheme in cyber-physical social networks. IEEE Trans. Comput. Soc. Syst. **5**(3), 773–784 (2018)

26. A. Jolfaei, K. Kant, H. Shafei, Secure data streaming to untrusted road side units in intelligent transportation system, in *18th IEEE International Conference On Trust, Security And Privacy In Computing And Communications/13th IEEE International Conference On Big Data Science And Engineering (TrustCom/BigDataSE)*. IEEE **2019**, pp. 793–798 (2019)

27. Y. Gong, C. Zhang, Y. Fang, J. Sun, Protecting location privacy for task allocation in ad hoc mobile cloud computing. IEEE Trans. Emerg. Topics Comput. **PP**(99), 1–1 (2016)

28. W. Wang, Q. Zhang, Privacy preservation for context sensing on smartphone. IEEE/ACM Trans. Netw. **24**(6), 3235–3247 (2016)

29. S.G. Ezabadi, A. Jolfaei, L. Kulik, R. Kotagiri, Differentially private streaming to untrusted edge servers in intelligent transportation system, in *2019 18th IEEE International Conference On Trust, Security And Privacy In Computing And Communications/13th IEEE International Conference On Big Data Science And Engineering (TrustCom/BigDataSE)*.IEEE, pp. 781–786 (2019)

30. A. Jolfaei, K. Kant, Privacy and security of connected vehicles in intelligent transportation system, in *2019 49th Annual IEEE/IFIP International Conference on Dependable Systems and Networks–Supplemental Volume (DSN-S)*. IEEE, pp. 9–10 (2019)

31. D.B. Rawat. K.Z. Ghafoor, *Smart Cities Cybersecurity and Privacy*. Elsevier (2018)

32. H. Zhang, Y. Shu, P. Cheng, J. Chen, Privacy and performance trade-off in cyber-physical systems. IEEE Netw. **30**(2), 62–66 (2016)

33. Y. Qu, L. Gao, T.H. Luan, Y. Xiang, S. Yu, B. Li, G. Zheng, Decentralized privacy using blockchain-enabled federated learning in fog computing. IEEE Int. Things J. (2020)

34. J. Yu, K. Wang, D. Zeng, C. Zhu, S. Guo, Privacy-preserving data aggregation computing in cyber-physical social systems. ACM Trans. Cyberphys. Syst. **3**(1), 8:1–8:23 (2019)

35. T. Um, G.M. Lee, J.K. Choi, Strengthening trust in the future social-cyber-physical infrastructure: an ITU-T perspective. IEEE Commun. Mag. **54**(9), 36–42 (2016)

36. I.J. Goodfellow, J. Pouget-Abadie, M. Mirza, B. Xu, D. Warde-Farley, S. Ozair, A.C. Courville, Y. Bengio, Generative adversarial nets, in *Advances in Neural Information Processing Systems 27: Annual Conference on Neural Information Processing Systems 2014, December 8-13 2014, Montreal, Quebec, Canada*, pp. 2672–2680 (2014)

37. C. Li, T. Xu, J. Zhu, B. Zhang, Triple generative adversarial nets, in *Advances in Neural Information Processing Systems 30: Annual Conference on Neural Information Processing Systems 2017, 4–9 December 2017, Long Beach, CA, USA*, pp. 4091–4101 (2017)

38. S. Song, K. Chaudhuri, A.D. Sarwate, Stochastic gradient descent with differentially private updates, in *IEEE Global Conference on Signal and Information Processing, GlobalSIP 2013, Austin, TX, USA, December 3–5, 2013*, pp. 245–248 (2013)

39. C. Dwork, V. Feldman, M. Hardt, T. Pitassi, O. Reingold, A. Roth, The reusable holdout: preserving validity in adaptive data analysis. Science **349**(6248), 636–638 (2015)

40. M. Abadi, A. Chu, I.J. Goodfellow, H.B. McMahan, I. Mironov, K. Talwar, L. Zhang, Deep learning with differential privacy, in *Proceedings of the 2016 ACM SIGSAC Conference on Computer and Communications Security, Vienna, Austria, October 24–28, 2016*, pp. 308–318 (2016)

41. R. Shokri, V. Shmatikov, Privacy-preserving deep learning, in *Proceedings of the 22nd ACM SIGSAC Conference on Computer and Communications Security, Denver, CO, USA, October 12–16, 2015*, pp. 1310–1321 (2015)

42. Y. Qu, M.R. Nosouhi, L. Cui, S. Yu, Privacy preservation in smart cities, in *Smart Cities Cybersecurity and Privacy*. Elsevier, pp. 75–88 (2019)
43. Y. Qu, S. Yu, W. Zhou, S. Peng, G. Wang, K. Xiao, Privacy of things: emerging challenges and opportunities in wireless internet of things. IEEE Wirel. Commun. **25**(6), 91–97 (2018)
44. J. Yu, K. Wang, D. Zeng, C. Zhu, S. Guo, Privacy-preserving data aggregation computing in cyber-physical social systems. ACM Trans. Cyber-Phys. Syst. **3**(1), 1–23 (2018)
45. L. Cui, G. Xie, Y. Qu, L. Gao, Y. Yang, Security and privacy in smart cities: challenges and opportunities. IEEE Access **6**,46 134–46 145 (2018)
46. L. Cui, Y. Qu, L. Gao, G. Xie, S. Yu, Detecting false data attacks using machine learning techniques in smart grid: a survey, J. Netw. Comput. Appl., 102808 (2020)
47. L. Gao, T.H. Luan, B. Gu, Y. Qu, Y. Xiang, *Privacy-Preserving in Edge Computing* (Springer, Wireless Networks, 2021)
48. L. Gao, T.H. Luan, B. Gu, Y. Qu, Y. Xiang, Blockchain based decentralized privacy preserving in edge computing, in *Privacy-Preserving in Edge Computing*. Springer, pp. 83–109 (2021)
49. L. Gao, T.H. Luan, B. Gu, Y. Qu, Y. Xiang, Context-aware privacy preserving in edge computing, in *Privacy-Preserving in Edge Computing*. Springer, pp. 35–63 (2021)
50. L. Gao, T.H. Luan, B. Gu, Y. Qu, Y. Xiang, An introduction to edge computing, in *Privacy-Preserving in Edge Computing*. Springer, pp. 1–14 (2021)
51. L. Gao, T.H. Luan, B. Gu, Y. Qu, Y. Xiang, Privacy issues in edge computing, in *Privacy-Preserving in Edge Computing*. Springer, pp. 15–34 (2021)
52. Y. Qu, L. Gao, Y. Xiang, Blockchain-driven privacy-preserving machine learning, *Blockchains for Network Security: Principles, Technologies and Applications*, pp. 189–200 (2020)
53. Y. Qu, M.R. Nosouhi, L. Cui, S. Yu, Personalized privacy protection in big data
54. L. Gao, T.H. Luan, B. Gu, Y. Qu, Y. Xiang, Existing privacy protection solutions, in *Personalized Privacy Protection in Big Data*. Springer, pp. 5–13 (2021)
55. L. Gao, T.H. Luan, B. Gu, Y. Qu, Y. Xiang, Future research directions, in *Personalized Privacy Protection in Big Data*. Springer, pp. 131–136 (2021)
56. L. Gao, T.H. Luan, B. Gu, Y. Qu, Y. Xiang, Leading attacks in privacy protection domain, in *Personalized Privacy Protection in Big Data*. Springer, pp. 15–21 (2021)
57. L. Gao, T.H. Luan, B. Gu, Y. Qu, Y. Xiang, Personalized privacy protection solutions, in *Personalized Privacy Protection in Big Data*. Springer, pp. 23–130 (2021)
58. J. Yoon, J. Jordon, M. van der Schaar, PATE-GAN: Generating synthetic data with differential privacy guarantees, in *International Conference on Learning Representations* (2019)
59. Y. Center, Yelp dataset 8th round, https://www.yelp.com/dataset_challenge

Chapter 5
Hybrid Privacy Protection of IoT Using Reinforcement Learning

The smart mobile device, as an indispensable component of IoT, has been growing in volume and diversity in recent years. Along with this, cyber-physical social network (CPSN) experiences fast booming, in which users publish their posts or data for sharing. However, since the published data is usually public to all, adversaries can crawl data or launch attacks without much efforts. Existing research usually considers a static adversary where the attack is launched once to steal a type of sensitive information like identity or location. This is not practical in real-world scenarios. To release this assumption, we develop a hybrid privacy-preserving model that protects identity and location privacy at the same time against a dynamic adversary who actively launches attacks. In the proposed model, the privacy protection problem is also considered as a trade-off optimization model that users target on maximizing data utility with high-level privacy protection but adversaries try to achieve a opposite target. To make this happen, we model a multi-stage game built upon Markov Decision Process (GMDP). The user and the adversaries are regarded as two players (also known as parties) in this dynamic zero-sum game. The output of this game should be the optimal actions of users that an adversary cannot breach more privacy no matter how the actions changed. An improved reinforcement learning algorithm based on state-action-reward-state-action (SARSA) is developed, which reduces the cardinality from n to 2. This can significantly improve the convergence efficiency. At last, we conduct experimental simulations on real-world datasets to testify the superior performance on efficiency and feasibility over existing research. This paper is mainly based on our research on privacy protection using reinforcement learning [1–5].

Y. Qu et al., *Privacy Preservation in IoT: Machine Learning Approaches*,
SpringerBriefs in Computer Science, https://doi.org/10.1007/978-981-19-1797-4_5

5.1 Overview

It is the ubiquitous existence of mobile devices and internet access that accelerates the popularization of the cyber-physical social networks (CPSN). The CPSN is an enhanced version of the classic social networks with the feature that the users actively publish data including sensitive information on the service apps [6]. Everyone can browse the location or nickname information, for example, a local business service system like "Groupon" or "Scoopon" [7]. In a CSPN, users act like sensors themselves and the data published is regarded as the sensing data [8], which is accessible to the public without access control.

Inappropriate release of sensitive information leads to privacy leakage and possible further damage. Although it is widely convinced that these applications can improve the quality of service and users can benefit from the released data [8], the possible damages caused by privacy leakage in CPSN have been widely considered by not only the government but also the public in recent years [9]. In CPSNs, both location privacy [10] and identity privacy [11] are under threats. These privacy issues are becoming severely afflicted areas. Firstly, the user publishes data with nickname or location to maximize his or her influence [12]. Secondly, the service provider indicates location or nickname of the user in various ways to maximize the service quality and user experience [13]. Thirdly, the data are set public and the adversary can easily craw them from the app without advanced hacker technology.

The target of privacy protection is to obtain an optimized trade-off between privacy and data utility [14]. Thus, privacy protection schemes also need to take data utility into consideration. Users may publish the data in a anonymous way to avoid privacy leakage that leads to poor data utility. We have an observation that current service providers usually cannot provide long-term multi-stage privacy protection [8]. Therefore, some of the users choose not to publish their location or identity information, which results in degradation of data utility.

Existing works, however, can not perfectly the new requirements of the CPSN. A lot of methods have been proposed to balance data utility and privacy protection [15], but most of previous works focus on either location privacy or identity privacy, for example, location coarsening, anonymity [16], k-anonymity, differential privacy [17], and so on. In the case of location coarsening and blocking technology, they use suppression and generalization methods to hide the location information, replace the GPS information with another adjacent one, produce pseudonyms or generalize the location into a coarse-grained area [18]. The k-anonymity-based methods are not applicable due to the published data is not easily clustered. Differential privacy-based methods also can not function well due to the same reason, in which the sensitive information is exclusive and distinguishable [19]. Although perturbation of the time slot is feasible and the data utility for other users is acceptable, the data utility for the user who publishes the data is damaged (e.g. most users are reluctant to publish data with inaccurate date). In [20], the authors proposed an optimal user-centric location-privacy preserving mechanism that provided good privacy protection, but they did not take the correlation and long-term attack into consideration. In addition, Mao et al. proposed a trust level system which can be used to establishment privacy preservation model [21].

In order to obtain the trade-off and derive the best strategy of the user, we model the confrontation between users and the adversary as a game-based Markov Decision Process (GMDP). We consider an adversary who can change strategies according to the situation. In the proposed model, the action of the user is regarded as the granularity of the sensitive contents, while the action of the adversary is modeled as the probability to eavesdrop on the information. The dynamic multi-stage zero-sum game lasts for finite stages, where the adversary and user keep adjusting their strategies to target on the best strategy of the user. The best strategy is proved to be the exclusive Nash Equilibrium (NE) point of this game. In order to gain the NE point fast and accurately, we introduce a modified state-action-reward-state-action (SARSA) reinforcement learning algorithm to minimize the convergence time. The computational complexity decreases sharply by reducing the cardinality from n to 2. Our theoretical analysis and extensive experiments indicate the effectiveness of the derived strategy.

We summarize the our contributions as below.

- We propose a hybrid privacy protection scheme to preserve both location privacy and identity privacy in cyber-physical social networks while existing works only focus on one of them. This is an early work to consider both privacy issues and take dynamic feature and multi-stage into account to provide hybrid and long-term privacy protection.
- We establish a mathematical model for the studied issue using a game-based Markov decision process, which is a dynamic multi-stage zero-sum game in essence. We deduce the best strategy of users by means of deriving the exclusive Nash Equilibrium.
- We implement a modified SARSA reinforcement learning algorithm to fast derive the exclusive Nash Equilibrium. The proposed algorithm is designed to deal with the state-action-reward-state-action problem, which captures the features of the game-based Markov decision process model. The modified SARSA algorithm can achieve a fast convergence by reducing the cardinality of the system states from n to 2.

The rest of this chapter is organized as below. In Sects. 5.2 and 5.3, we present the preliminaries and problem statement, respectively. We then present framework of the GMDP model in Sect. 5.4. Section 5.5 depicts the system analysis, which is followed by the modified SARSA algorithm to derive the best strategy in Sect. 5.6. The performance and evaluation are illustrated in Sect. 5.7. At last, Sect. 5.8 summarizes and concludes this chapter.

5.2   Related Work

There are extensive existing privacy-preserving schemes which provide strong privacy protection in certain perspective. There are two directions in privacy-preserving data publishing field. The first one is clustering-based methods, for example, K-anonymity [22], L-diversity [23], T-closeness [24], and their variants [25–27]. The other one is differential privacy [28] and its variants. The clustering-based methods have been proved effective and feasible. However, there is no solid theoretical foundation and they cannot meet the requirements of big datasets. The differential privacy and its variants have been widely used in variant scenarios. They have solid theoretical foundation and provide high-level privacy protection in data retrieval scenarios.

The models mentioned above have also been modified and introduced into location privacy protection. There are dummy-based method [18], spatial cloaking [15], and anonymization [29], which are widely used in current research works. The dummy-based methods generate dummies and send it to the service provider after filtering. The anonymization methods mainly use pseudonyms and change them over time. Location privacy protections for ad hoc networks, such as, the mobile networks [30], and vehicular networks [31, 32], are taken into consideration and have promoted in recent years. Personalized privacy is a new topic and widely researched these years. However, it suffers form inference attack [33]. Koufogiannis et al. proposed an idea of protecting customized privacy based on the social distance of a social network [34]. These studies have focused on single-shot location privacy, while the published data of multi-stage is not considered. There are some existing works protection privacy with game theory [35]. In [8], the authors proposed the idea to model the privacy issues in context sensing scenario as a competitive MDP. In [36], Zhang et al. proposed a personalized spam filtering model with privacy protection.

Interested readers are encouraged to explore literature review we have done in recent years [37–41]. Besides, we also summarize several of our book collections to show the frontier of privacy preservation research in edge computing [42–47] and privacy preservation research in edge computing in big data [48–52].

In addition, the authors did deep research on Markov Decision Process in [53]. Both theory and applications are deeply discussed. Leslie et al. [54] did a comparison between different reinforcement learning methods and illustrated the advantages of SARSA.

5.3   Hybrid Privacy Problem Formulation

5.3.1   Game-Based Markov Decision Process

A game-based MDP is a dynamic zero-sum game with multiple finite stages, in which there is probabilistic transactions. In this scenario, there are two players and we use a seven-tuple, $\{S_1, S_2, A_1, A_2, R_1, R_2, Pr\}$, to denote the **Game-based MDP**. S_j is the discrete space of player j where $j = 1, 2$. A_j is the action space of player j where $j = 1, 2$. $R_j: S_1 \times S_2 \times A_1 \times A_2 \longmapsto R$ is the payoff function of player j, in which $\longmapsto$ means the mapping of input and output. $Pr: S_1 \times S_2 \times A_1 \times A_2 \longmapsto \delta S$ is the transition probability map, where $\delta(S_1, S_2)$ is the set of probability distributions over the state pair (S_1, S_2).

5.3.2 Problem Formulation

In cyber-physical social networks, service suppliers provide all kinds of services, for example, commenting on restaurants in local location-based systems. To start with, we split the whole region into many disjoint unit regions $r = \{r_1, r_2, ...r_n\}$. Every unit region corresponds to a geographically closing district, like a block or a shopping center. Each service only locates in one unit region. Users are nodes of the system who would like to publish some information and location relevant to the unit region they are situated. Users are described as $U = \{u_1, u_2, ..., u_n\}$. We use message to denote the published data of the user, where $M = \{m_1, m_2, ..., m_n\}$. For clarity, we model the time into a series of time slots and use stage $T = \{t_1, t_2, ..., t_n\}$ to describe it. In the stages that the user publishes a message, the state is $\{r_i, u_i, m_i, t_i\}$.

The initial stage is $t = 0$ when the life cycle of the service starts. The adversary starts to attack at a certain stage and tries to eavesdrop all the existing messages of the user inside the system. After the attack, the user changes the strategy to defend the adversary according to the attack result of last stage. In the following stages, the user and adversary starts the confrontation in a game-based MDP. The game is

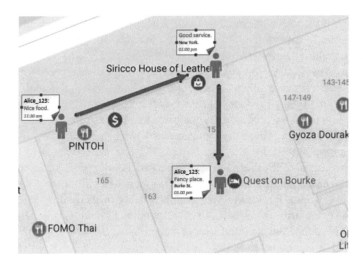

Fig. 5.1 An example of private data publishing in IoTs

played in finite stages and the user targets on acquiring the best strategy to maximize the overall payoff. For the sake of avoiding being re-identified by the adversary, we assume that the user publishes the location or nickname, and even real name. For instance, in Fig. 5.1, a user "Alice" follows a trace from a restaurant to a store and arrives at a hotel as the terminal in different time slots. Alice publishes a message with location, identity, and both of them, respectively.

5.4 System Modelling

In this section, we establish the proposed model to provide protection for both identity and location privacy. We firstly describe the basic models proposed in GMDP, including actions of the adversary and user, and system states transmissions. Meanwhile, we introduce unit region and define two privacy metrics to measure both location privacy and identity privacy. At last, the problem is formulated into a GMDP, in which the user wants to maximize the overall payoff while the adversary targets on minimizing it. In this model, we take both location and identity into consideration to provide comprehensive and high-level privacy protection for CPSN.

5.4.1 Actions of the Adversary and User

For the attack, the adversary collects instant messages of the user positively in order to infer the location or identity. According to the sensitive contents of the messages, the adversary can conduct attack including pushing spams to the user, stealing private information of bank accounts, or even committing physical crime.

In this subsection, we discuss a dynamic adversary AD, which has limited computing abilities to eavesdrop and analyze the messages of the user. This is different from the popular adversary model named global adversary who has the unlimited computing power to traverse a system. However, this global adversary is not feasible and loses focus on real problems. Therefore, we assume this "intelligent" adversary AD can not modify any messages or inject some. AD actively collects the information according to the M published by the users and tempts to breach the privacy of location and identity, while the user defenses the adversary by changing his strategies through multiple stages.

The action serial of the user is defined as the granularity of the released message. After taking the last state S_{t-1} into account, the user decides his next action (maintain the same or change). For clarity, we use the veracity of the published message to measure the granularity, which is $\{0, 1\}$ corresponding to anonymity and full-release. Therefore, we define the action of user at stage t as A_u^t, which is the granularity of how much information is released.

The actions serial of the adversary is regarded as the probability of whether to eavesdrop. The adversary has limited computing resources to observe the released messages of the user. The adversary also has an interest which impacts the probability to eavesdrop. Formally, we define the action of the adversary at stage t as A_{ad}^t, which denotes the probability of whether the adversary takes action. Furthermore, we formulate the computing constrains of the adversary as followings.

$$\sum_t A_{ad}^t \leq \Psi, \ 0 \leq A_{ad}^t \leq 1, \tag{5.1}$$

where the Ψ is the maximum computing power of the adversary. When $\Psi \geq t$, the adversary has the ability of eavesdropping all the messages in each stage while $\Psi < t$ means the computing ability of the adversary is limited.

5.4.2 System States and Transitions

5.4.2.1 System States

In this system, the user acts as a sensor himself and release data (message) to public for further utility. The messages of the user are published in the system as the action of the user depends on his observation of the current states. The adversary could infer the location or identity of the user and even the Markov state transitions as well.

For the attack, we should consider both the current message and action of the adversary. It is obvious that a user has no clue of the adversary's action or strategy. Therefore, a user can only reckon the action of the adversary from a former attack output, which is easily observed by the user. On one hand, the former attack result shows if an adversary successfully breach the privacy and infer the information. On the other hand, the current state has one and only one precondition, which is the last state in a MDP. For example, if a user receives a notification service which is based on his current location, he is sure that the adversary has already successfully breached the location and identity privacy, while if the user receives a notification service based on a wrong or previous message, the user is sure of the adversary's failure. That is the reason why the user should save a record of messages which the adversary has targeted on and, in which the adversary has successfully breached. We use AR_t to denote the attack result which is observed at stage t. If $AR_t = 1$, we say the adversary successfully obtains sensitive messages in M_{t-1}, while $AR_t = 0.5$ describes partial re-identification and $AR_t = 0$ means the failure of the adversary. All in all, the current action of the user is determined by the message of current stage and attack result of last stage. Therefore, we define the the system state of attack as Eq. (5.2).

$$S^t = \{M_t, AR_t\} \tag{5.2}$$

5.4.2.2 State Transitions of the System

In this subsection, we discuss the state transitions in attack. The non determinacy of state (S_1^t, S_2^t) results from the uncertainty of the M_t, while the state is determined by the actions of both the adversary and the user. At the same time, the attack result AR_t is determined by the action as well. Therefore, we define the state transition in attack as

$$\Pr\left[(S_u^t, S_{ad}^t) \Big| (S_1^{t-1}, S_1^{t-1}), A_u^{t-1}, A_{ad}^{t-1}\right]$$
$$= \Pr\left[M_t \Big| M_{t-1}\right] \Pr\left[AR^t \Big| AR^{t-1}, A_u^{t-1}, A_{ad}^{t-1}\right] \qquad (5.3)$$
$$= \Pr\left[M_t \Big| M_{t-1}\right] \Pr\left[AR^t \Big| A_u^{t-1}, A_{ad}^{t-1}\right].$$

We gain the second equality because the attack result AR^t at stage t only depends on the actions of user and adversary at stage $t-1$. Thus, AR^t is not impacted by AR^{t-1}.

5.4.3 Nash Equilibrium Under Game-Based MDP

We observe that the payoff function and the transition map are both based on states and actions of all players. The game is played in multiple finite stages and for each stage there is a payoff $r_j(S_u, S_{ad}, A_u, A_{ad})$ obtained from actions of players $a_j \in A_j$ and current states $s_j \in S_j$. Every user j has a goal of maximizing overall discounted payoff while the adversary targets on minimizing it. Therefore, we formulate the problem as deriving the NE of the multi-stage zero-sum dynamic game in a game-based MDP.

According to our analysis, the payoff of each stage is based on the states and actions of users and the adversary. We define the payoff function as the difference of QoS and the privacy loss, which can be expressed as

$$R\left((S_u^t, S_{ad}^t), A_u^t, A_{ad}^t\right) = QoS\left(A_u^t\right) - \omega \cdot PL\left(S_u^t, S_{ad}^t\right), \qquad (5.4)$$

where $QoS(A_u^t)$ is the quality of service that the user experiences, $PL(S_u^t, S_{ad}^t)$ is the privacy loss of the m_t, and ω is the quality of service decrease per unit privacy loss.

In a dynamic multi-stage zero-sum game, we define a strategy to be probabilistic transitions over the action sets of players. We also define a stationary strategy τ, which maintains the same ($\tau_t = \tau$) for all stage t. We discuss the case of stationary strategy in our model, which is more feasible in real-world scenario. In this dynamic multi-stage zero-sum game, we use $\tau^u : S_u \longmapsto \delta(A_u)$ to denote the strategy of the user, and $\tau^{ad} : S_{ad} \longmapsto \delta(A_{ad})$ to denote the strategy of the adversary, where (S_u, S_{ad})

denotes the state space, $\delta(A_u)$ and $\delta(A_{ad})$ the probability distribution over A_u and A_{ad} which are the action spaces of the user and the adversary.

In this case, we define a initial state when the user first publishes a message after registration. We regard the state $t = 0$ as the initial stage and the initial state is expressed as S_0. Given a specific state $s \in S$ and strategies of the user and the adversary, we rewrite the utility of the user at stage t as

$$PU^{\tau}(s) = \sum_{t=0} \mathrm{E}\left[R\left((S_u^t, S_{ad}^t), A_u^t, A_{ad}^t\right)\Big| \tau_u, \tau_{ad}, S^0 = s\right]. \tag{5.5}$$

As discussed above, we focus on stationary strategy which means that the actions A_u^t, A_{ad}^t are decided by the policies τ_u, τ_{ad}. Therefore, we can derive from Eq. (5.5) and get

$$PU^{\tau}(s) = R\left(s, A_u^t, A_{ad}^t\right) + \sum_{\hat{s}} \Pr\left[\hat{s}\big| s, A_s^t, A_{ad}^t\right] PU^{\tau}\left(\hat{s}\right). \tag{5.6}$$

As both the user and the adversary want to follow their best policies, we define τ_u^* and τ_{ad}^* as the best policies respectively. For a certain stage, we use a best strategy pair, which is $\tau^* = \{\tau_u^* + \tau_{ad}^*\}$, to represent the best strategies for current stage. According to game theory, we know that the best strategy pair is the NE for the dynamic multi-stage zero-sum game and we formulate it as followings.

[Nash Equilibrium]
Given a dynamic multi-stage zero-sum game and the state $s \in S$, the NE of it is the best strategy pair $\tau^* = \{\tau_u^* + \tau_{ad}^*\}$, we have

$$\begin{cases} PU^{\tau^*}(s) \geq PU^{\tau_{ad}}(s), \\ PU^{\tau^*}(s) \leq PU^{\tau_u}(s), \end{cases} \tag{5.7}$$

where $\tau_{ad} = \{\tau_u, \tau_{ad}^*\}$, $\tau_u = \{\tau_u^*, \tau_{ad}\}$, for all τ_{ad} and τ_u.

As the adversary wants to use his τ^{ad*} to get the minimum $PU^{\tau}(s)$ while the user wants to use τ^{u*} to protect his utility. According to Eq. (5.6), we have

$$PU^{\tau^*}(s) = \max_{\tau_u} \min_{\tau_{ad}} \left\{ R\left(s, A_u^t, A_{ad}^t\right) \right. \\ \left. + \sum_{\hat{s}} \Pr\left[\hat{s}\big| s, A_u^t, A_{ad}^t\right] PU^{\tau}(\hat{s}) \right\}. \tag{5.8}$$

In [55], the authors have proved that there is an exclusive NE in a dynamic multi-stage zero-sum game, which means the best strategy pair is the exclusive NE in our model.

5.5 System Analysis

5.5.1 Measurement of Overall Data Utility

In the proposed model, $QoS(a_u^t)$ denotes the satisfaction of the user when publishing the message m_i. Thus, we use the tanh function to model it. Tanh function is widely used to measure the satisfaction of user according to the quality of service. The reason why we use tanh function includes three aspects. The first one is that the satisfaction of the user maintains low if and only if the quality of service changes in a low range. The second one is that the satisfaction of the user grows rapidly if and only if the accuracy grows across a satisfaction threshold. The last one is that if the accuracy has already entered a high range, increasing improvements brings barely no further benefits to the user. Therefore, we formulate $QoS(A_u^t)$ as

$$QoS(A_u^t) = \frac{2\mathrm{Exp}\Big(-\rho(g-\sigma)\Big)}{1+\mathrm{Exp}\Big(-\rho(g-\sigma)\Big)} - 1, \tag{5.9}$$

where ρ determines the steepness of service quality curve, g the recognition accuracy, and σ is the threshold, below which the satisfaction of the user is very limited, and above which the satisfaction of the user grows quickly and achieves a convergence, while the curves before and after the threshold are convex and concave, respectively.

5.5.2 Measurement of Privacy Loss

To measure privacy loss, we define a user who has a message space M and a series of messages $M_s \in M$ with his location or identity information. The released data is regarded as privacy-preserving if and only if the adversary learns nothing or little information about the user and the adversary cannot infer any useful information about the user. For all the sensitive messages and stages, the differences between the adversary's prior and poster beliefs on the user should be properly limited. In real-world scenario, we have the observation that an adversary pay more attention on user's recent released messages rather than future messages. Therefore, we measure the privacy loss according to context privacy [56] and information Entropy.

   The sensitivity of a series of messages $m_t \in M$ is the sum of discounted differences which are between the prior belief and poster belief of the adversary eavesdropping the sensitive messages at the current stage, therefore, we have

$$MSen(m) = \sum_{t=0}\sum_{m_s \in M_s}\Big| \Pr\Big[m_t = m_s\Big|m_0 = m\Big] - \Pr\Big[m_t = m_s\Big]\Big|, \tag{5.10}$$

The sensitivity of a single message m_t denotes how much information that the adversary can obtain before and after eavesdropping the current m_t.

We further model privacy loss of user based on the message sensitivity. With regard to the attack, there are three attack results. For the case of successfully inferring the location and identity privacy information, the privacy loss is the message sensitivity. If the adversary only successfully infers one of identity or location privacy information, the privacy loss is message sensitivity multiplied by a impact factor. But if the adversary fails to infer anything, we can conclude that the privacy is zero because the adversary gains no extra information from the current message. Therefore, we formulate the privacy loss in attack as

$$PL\left((S_u^t, S_{ad}^t), A_u^t, A_{ad}^t\right) = MSen\left(M_t\right)AR_{t+1}, \qquad (5.11)$$

where AR^{t+1} is the attack result at the stage $t + 1$. And it is intuitive to know that the probability of successfully inferring a message at stage t is $Pr[AR_{t+1}] = A_u^t \cdot A_{ad}^t$.

After defining the quality of service and the privacy loss attack, we discuss about ω, which denotes the service quality change based on unit privacy loss. For each message, we measure all the privacy loss and QoS improvement when the adversary can eavesdrop all the information of the user, compared to the situation that the adversary fails to eavesdrop any information from the user. In this work, we assume the common case that the adversary has some background knowledge to infer the sensitive information according to the messages of the user. For example, the adversary may have a clue of the user's approximate pattern of behaviors or previous Markov state transitions. After taking Eqs. (5.9)–(5.11) into (5.4), we can calculate the stage payoff of the user. At the same time, we can obtain the stage payoff of the adversary is the minus of the user's.

Normally, the user keeps updating the messages during the lifecycle of the service. Therefore, it is intuitive to conclude that the game-based MDP is played in infinite stages. As the discussion above, the adversary values more on current message rather than future messages because the adversary could launch attacks according to current sensitive information. In this case, we formulate the utility of the user as the expected sum of payoffs in each stage. Generally, the user pays more attention to future payoffs. Thus, the utility of the user can be expressed as

$$U_u = E\left[\sum_{t=0} R\left((S_u^t, S_{ad}^t), A_u^t, A_{ad}^t\right)\right], \qquad (5.12)$$

In this game-based MDP, the goal of the user is to find out the best strategy (series of actions) to defense the adversary, which refers to find out the maximum U_u in our model, while the adversary has the opposite goal, which is to gain maximum privacy information and to benefit from it. Thus, the adversary aims at minimizing U_u.

5.6 Markov Decision Process and Reinforcement Learning

We make modification to the SARSA reinforcement learning algorithm to tackle the problem of deriving the exclusive NE point of the zero-sum multi-stage game in game-based MDP. The proposed method lessens the computational complexity by reducing the cardinality from n to 2. Moreover, we use the NE point to obtain the best strategy of the user when the adversary has limited computing ability. In addition, an extension is made to fit for the situation that the adversary has unlimited computing ability.

5.6.1 Quick-Convergent Reinforcement Learning Algorithm

In order to derive the NE point of the dynamic multi-stage zero-sum game, we introduce reinforcement learning into our model. However, existing models are not perfect suitable to this scenario. For example, the temporal difference algorithm needs a model, the mc-control model considers the cardinality of the states, which is a large number and significantly affects the computing complexity and further leads to adverse effects to time complexity. In our study, we use a modified SARSA algorithm to quickly derive the NE point by solving an equivalent issue. We define the equivalent $\widehat{PU}_u^{\tau^*}(AR)$ as the expected of $PU_u^{\tau^*}(s)$ where $s = \{AR, M\}$, through which we can get rid of M. We express the $\widehat{PU}_u^{\tau^*}(AR)$ as

$$\widehat{PU}_u^{\tau^*}(AR) = E\left[R_u\left(s, A^{\tau^*}\right) + \sum_{AR'} \left(\Pr\left[AR' \middle| A^{\tau^*} \right] \widehat{PU}^{\tau^*}\left(AR' \right) \right) \right], \tag{5.13}$$

where $A^{\tau^*} = \{A_u^{\tau^*}, A_{ad}^{\tau^*}\}$ is the best action following the best strategy τ^*.

According to Eq. (5.13), we reduce the cardinality of $|S|$ into 3 as the AR has only 3 values. Therefore, we can obtain the best strategy pair τ^* by deriving an equivalent problem as followings.

$$\tau^* = \max_{\tau_u} \min_{\tau_{ad}} E\left[R_u\left(s, A^{\tau^*}\right) + \sum_{AR'} \left(\Pr\left[AR' \middle| A^{\tau^*} \right] \widehat{PU}^{\tau^*}\left(AR' \right) \right) \right] \tag{5.14}$$

Based on Eq. (5.14), we can use $\widehat{PU}_u^{\tau^*}(AR)$ to derive the best strategy τ^*. We use the following updating-rule to derive $\widehat{PU}_u^{\tau^*}(AR)$, which is modified from the classic SARSA algorithm.

$$
\widehat{PU}^{t+1}(AR) = \left(1 - \alpha_{t+1}\right)\widehat{PU}^t(AR) +
$$
$$
\alpha_{t+1}\mathrm{E}\left[R\left(s, A_u^t, A_{ad}^t\right) + \widehat{PU}^t(AR')\right], \tag{5.15}
$$

where $\alpha_t \in [0, 1]$ denotes the learning rate of the algorithm. We set α_t to decrease with time in order to get a deterministic convergence, which is $\alpha_t = 1/t$. In this update step, $\widehat{PU}^{t+1}(AR)$ is regarded as the approximate value of $\widehat{PU}^{\tau^*}(AR)$ and it finally converges to $\widehat{PU}^{\tau^*}(AR)$ after finite rounds of updates.

In a classic SARSA reinforcement learning method, the computational complexity is $O(|A|^3|S|^3)$, in which, S means the states set and A denotes the action set. In the proposed method, we have proved the cardinality of action reduces from n to 2. Therefore, the computational complexity of this algorithm changes to $O(2^3|S|^3)$. In a Markov Decision Process, the cardinality of action A is normally large and thus the reduction of cardinality results in significant performance up-gradation.

5.6.2 Best Strategy Generation with Limited Power

In this model, we need to derive the Nash Equilibrium of the multi-stage game to gain the $\widehat{PU}^t(AR)$, which denotes the state value instead of $\widehat{PU}^t(s)$. We can derive the NE from Eq. (5.14) and the game value from Eq. (5.15). Through submitting Eqs. (5.1), (5.4), (5.9)–(5.11) into (5.14), we reshape the NE-derivation problem as followings.

$$
\min_{\tau_{ad}} \max_{\tau_u} \left\{ \frac{2\mathrm{Exp}\left(-\rho(A_u^t - \sigma)\right)}{1 + \mathrm{Exp}\left(-\rho(A_u^t - \sigma)\right)} - M(m)A_{ad}^t A_u^t - 1 \right\},
$$
$$
s.t.
$$
$$
\sum_t A_{ad}^t \leq \Psi, \tag{5.16}
$$
$$
0 \leq A_{ad}^t \leq 1, \forall t,
$$
$$
0 \leq A_u^t \leq 1, \forall t,
$$

where $M(m)$ is the function of the message m that $M(m) = \omega M\,Sen(m) + (\widehat{PU}^{\tau^*}(AR' = 0) - \widehat{PU}^{\tau^*}(AR' = 1))$. As the $\tau^*(AR' = 0)$ and $\widehat{PU}^{\tau^*}(AR' = 1)$ maintains the same, which means that $M(m)$ is only depends on m. In above analysis, we have the observation that $\widehat{PU}^{\tau^*}(AR' = 0) > \widehat{PU}^{\tau^*}(AR' = 1)$. As $MSen \geq 0$, we conclude that the function of message $M(m) > 0$.

For the sake of solving Eq. (5.16), we first eliminate the effects of the adversary. As we focus on stationary strategy in this chapter. τ_u is fixed to a constant value. As we have proved $M(m) > 0$, we assume the adversary to eavesdrop Ψ messages to launch the attack to minimize the value in Eq. (5.16). Therefore, we re-formulate the problem as

$$\max_{\tau_u, \Theta, T'} \left\{ \frac{2\mathrm{Exp}\left(-\rho(a_u^t - \sigma)\right)}{1 + \mathrm{Exp}\left(-\rho(A_u^t - \sigma)\right)} - M(m)A_u^t - 1 \right\},$$

$$s.t. \tag{5.17}$$

$$0 \leq A_u^t \leq 1, \forall t,$$
$$A_{ad}^t \leq \Theta, \forall t \in T',$$
$$A_u^t \geq \Theta, \forall t \in \{T/T', \}$$

where T' is one subset of T that is consist of Θ messages. Given a certain T', we can easily derive the closed-form of best strategy τ_u.

5.6.3   Best Strategy Generation with Unlimited Power

In the previous subsections, we assume the adversary has limited computing power and can only eavesdrops the messages with his best strategy. Whereas, we focus on the adversary with unlimited computing power, who can access all the messages the user publishes. Although this is a different scenario, the proposed learning algorithm is still suitable and we continue to use the modified SARSA algorithm to derive the NE point in the situation of unlimited computing power. Therefore, we have

$$\max_{\tau_u} \left\{ \frac{2\mathrm{Exp}\left(-\rho(A_u^t - \sigma)\right)}{1 + \mathrm{Exp}\left(-\rho(A_u^t - \sigma)\right)} - M(m)max_{\tau^*}A_{ad}^t A_u^t - 1 \right\},$$

$$s.t. \tag{5.18}$$

$$0 \leq A_{ad}^t \leq 1, \forall t,$$
$$0 \leq A_u^t \leq 1, \forall t,$$

where the constrain of computing power is eliminated from the equation. Intuitively, we can observe that $max_{\tau^*}A_{ad}^t A_u^t = A_u^t$. Thus, we can re-formulate Eq. (5.18) into

$$\max_{\tau_u} \left\{ \frac{2\text{Exp}\Big(-\rho(A_u^t - \sigma)\Big)}{1 + \text{Exp}\Big(-\rho(A_u^t - \sigma)\Big)} - M(m)A_u^t - 1 \right\},$$

$$s.t. \tag{5.19}$$

$$0 \le A_{ad}^t \le 1, \forall t,$$
$$0 \le A_u^t \le 1, \forall t,$$

where we can derive the closed-form expression of the best strategy τ_u.

Algorithm 4 Quick-Convergence Learning Algorithm

Input: The multi-stage zero-sum dynamic game;
Output: The best stratety π^*;
1: $t = 0$, $AR_{t=0} = 0$;
2: $\widehat{PU}^t(AR = 0) = 1$, $\widehat{PU}^t(AR = 0.5) = 1$, $\widehat{PU}^t(AR = 1) = 1$;
3: Initialize the strategy pair $\{\pi_u, \pi_{ad}\}$;
4: **while** Until Converge **do**
5:    Select current action pair based on $\{\pi_u, \pi_{ad}\}$;
6:    Update value of AR^{t+1} after taking actions;
7:    Update state value $\widehat{PU}^t(AR)$;
8:    Update the best strategy π_{t+1} with $\widehat{PU}^t(AR)$;
9:    Next stage: $t = t + 1$;
10: **end while**

5.7 Performance Evaluation

In this section, we first testify the effectiveness of our model based on measurements of data utility and privacy loss with different arguments, and further prove the rapid convergence of the proposed algorithm. We evaluate our model on the 8th round Yelp dataset [57] released by Yelp Inc. In this dataset, we mainly focus on the information of business, user, and the reviews, which correspond to location, identity and user action in our model, respectively. We describe user's action according to this real-world dataset and simulate user's action according to the model we proposed in Sect. 5.4. The algorithms are implemented on Java and are executed on Mac OS platform with Core I5@2.7 GHz CPU and 8G memory.

The details of the 8th round Yelp dataset is as followings. The data is from 11 cities in 4 different countries across Europe and North America. There are 4.1M reviews and 947 K tips by 1M users for 144 K businesses. In the business part, there are totally 1.1M business attributes, e.g., hours, parking availability, and ambience.

In the following experiments, we compare the proposed algorithm with two baseline algorithms, which are the constant strategy and the shortsighted strategy. In the constant strategy, the user sets the same granularity $(1/n)$ for all actions, while in the shortsighted strategy, the user does not care about the payoff in the future and tries to maximize the current payoff.

5.7.1 Experiments Foundations

In the case of system parameters, we have the following initial settings unless there are specific statements. The parameter of satisfactory is set to 8. We also set number of the messages to re-identify the location or identity privacy is 5. The weights of each message follows the uniform distribution.

Based on the parameters, we have the early-stage configurations as below.

- Firstly, we use a random algorithm to choose 10,000 records with valid ID, GPS, and reviews information from the Yelp dataset;
- Secondly, we use the top 4000 records as a training dataset for reinforcement learning of Markov Decision Process in both classic SARSA and the modified SARSA algorithms;
- Thirdly, the proposed algorithm and two baseline algorithms are performed on the trained GMDP model to compare the privacy level, data utility, and iteration times;
- Lastly, the proposed algorithm are implemented on the classic SARSA and the modified SARSA algorithms to illustrate the advantage of time consumption through cardinality reduction.

5.7.2 Data Utility Evaluations

In Figs. 5.2 and 5.3, we compare the performances of the three algorithms when facing the same dynamic adversary in an attack with limited and unlimited computing ability correspondingly. It is shown that in all three cities, the proposed algorithm and the shortsighted algorithm have better performances than the static algorithm. That's because the static algorithm never changes the actions and does not change strategy with the dynamic adversary.

When comparing the proposed algorithm and the shortsighted algorithm, we can see that the proposed algorithm overweights the shortsighted algorithm. The reason why the proposed algorithm achieves a higher payoff is that the shortsighted algorithm does not consider the future payoff as entirety. It only focus on maximizing the current payoff, which makes it unable to take the overall payoff into consideration.

Furthermore, by comparing Figs. 5.2 and 5.3, we can conclude the payoff of the user will decrease with the increase of the adversary's computing ability and a sharp decrease occurred when the adversary achieves the status of unlimited computing ability.

In Fig. 5.4, we can see that the payoff decreases with the increase of the percentage of the sensitive messages in the proposed algorithm and the shortsighted algorithm,

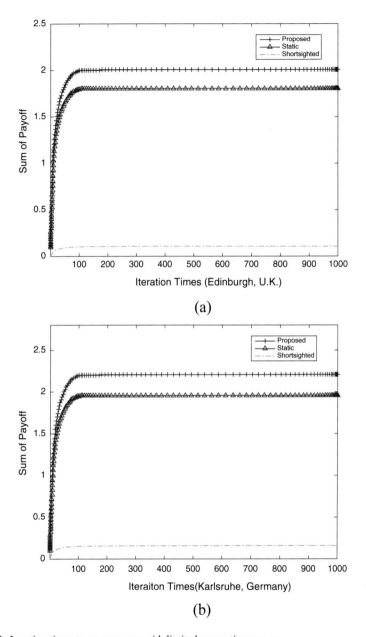

Fig. 5.2 Iteration times to convergence with limited computing power

while it maintains the same in a low level in the static algorithm. In the case of the static algorithm, the service quality maintains the same regardless of the adversary's actions, which leads to the fixed payoff. However, the reason why payoff decreases

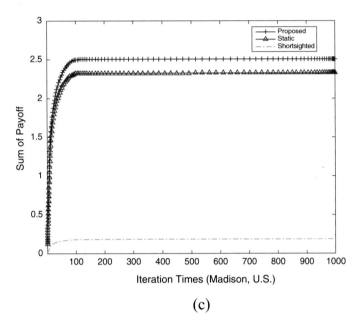

(c)

Fig. 5.2 (continued)

in other two algorithms is that there is more privacy loss when the sensitive messages has a higher percentage. Moreover, the figure depicts that the difference between the proposed algorithm and the shortsighted algorithm is smaller when the percentage is small than the difference when the percentage is large. This indicates when there is less sensitive information in the messages, the user's action has less impact on the future payoff.

Figure 5.5 describes the trend of the payoff with the change of the satisfactory threshold. All three algorithms have relative high payoffs with a small satisfactory threshold. That's because the $QoS(s)$ is large with a small satisfactory threshold. However, with the increase of the satisfactory threshold, payoffs of all three algorithms decrease and payoff of the static algorithm achieves almost 0 when satisfactory threshold is 1. The proposed algorithm has a better performance than the shortsighted algorithm in term of the service quality.

Furthermore, from Figs. 5.4 and 5.5, we can conclude that the mechanism can ignore the impact of future payoff if the messages have a small percentage of sensitive information while the user has a low satisfactory degree. For most real-world scenarios, either the percentage or the threshold, or even both of them is relative large, which make our proposed algorithm practical and feasible.

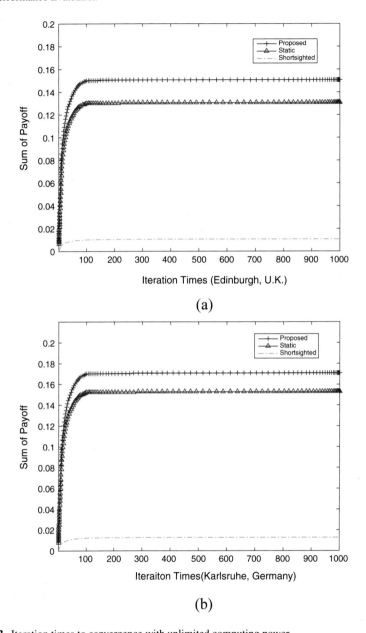

Fig. 5.3 Iteration times to convergence with unlimited computing power

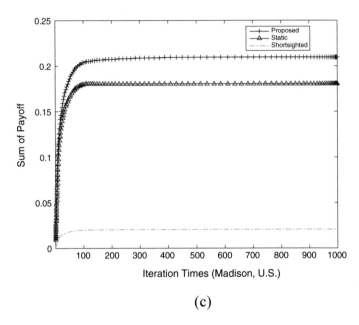

(c)

Fig. 5.3 (continued)

5.7.3   Privacy Loss Evaluations

In order to prevent the track from the adversary, the user can choose anonymity, partial release, or full release to publish the messages. We split the whole area into unit zones and each message can only fall into one of the unit zones.

5.7.3.1   Density of Location and Identity

For each unit region, we have a specific assumption of the adversary's knowledge as followings. The adversary can obtain the accurate statistics of the number of public reviews and locations (the number may not be the same for not all of the users publish both information). As a matter of fact, this kind of information can be provided by the service provider or an adversary can even crawl the sensitive data from the websites himself. According to these data, we formalize the location density and identity density for each of unit regions, respectively. The location density and identity density are defined as Eqs. (5.20) and (5.21).

$$\eta_i = \frac{n_i^L}{N_L}, \tag{5.20}$$

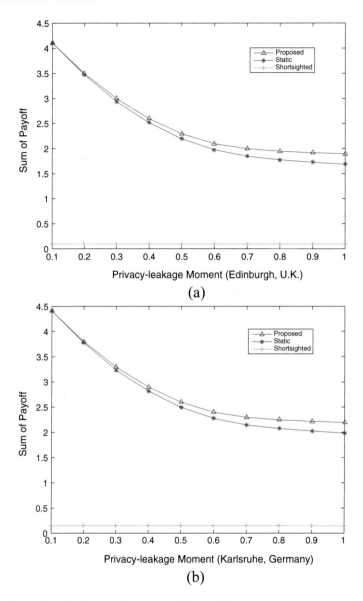

Fig. 5.4 Sum of payoff changes with increase of the sensitive messages

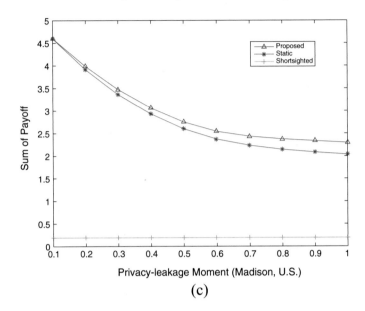

Fig. 5.4 (continued)

$$\zeta_i = \frac{n_i^{ID}}{N_{ID}}, \tag{5.21}$$

where n_i^{ID} is the number of users publishing their identity information (e.g. a nick-name) at unit region r_i, and n_i^L is the number of users publishing their location information (e.g. a restaurant or a cinema) at unit region r_i, while N_{ID} is the number of total users and N_L is the number of services in r_i.

5.7.3.2 Privacy Metrics of Location and Identity

We quantify the location privacy and the identity privacy provided by unit zone when there is an adversary eavesdropping on the data. We measure the privacy loss of the user with the entropy metric from the information theory. To generalize entropy in an unit region, the normalized entropy is computed for each service (location) k_L and each user k_{ID}, after which we get the weighted sum over all the possibilities based on the location density η_{k_L} and identity density $\zeta_{k_{ID}}$.

In the case of location privacy, we divide the result by η_{k_L} to get the normalized entropy, which is E_i^L at unit region r_i as Eq. (5.22).

$$E_i^L = -\sum_{k_L} \eta_{k_L} \Pr\left[k = k_L\right] \log_2 \Pr\left[k = k_L\right], \tag{5.22}$$

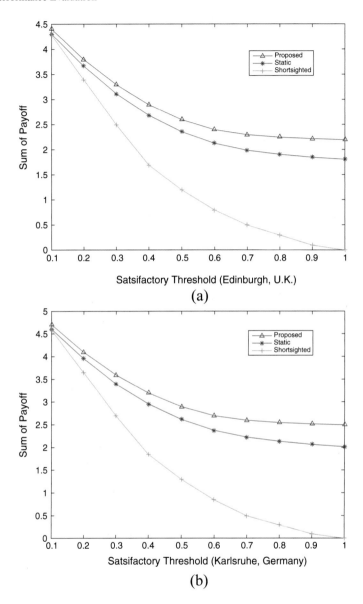

Fig. 5.5 Sum of payoff changes with increase of the satisfactory threshold

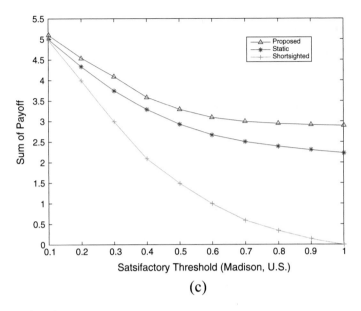

(c)

Fig. 5.5 (continued)

where $\Pr\left[k = k_L\right]$ is the possibility that an user publishes a message with location information of service k_L. The entropy E_i^L depicts the uncertainty of the adversary about the different services inside an unit region.

As to identity privacy, we divide the result by ζ_{k_ID} to get the normalized entropy, which is $E^I D_i$ at unit region r_i, as Eq. (5.23).

$$E_i^{ID} = -\sum_{k_{ID}} \zeta_{k_{ID}} \Pr\left[k = k_{ID}\right] \log_2 \Pr\left[k = k_{ID}\right], \qquad (5.23)$$

where $p_{k_{ID}}$ is the possibility that an user publishes a message with identity information of identity density k_{ID}. The entropy E_i^{ID} depicts the uncertainty of the adversary about the user inside an unit region.

It is intuitive to observe that re-identifying or re-location becomes more difficult for an adversary if there is a greater number of users publishing their messages within an unit region, while the uncertainty keeps increasing with the increase of the data sizes. Therefore, we can conclude that the mixing effectiveness at unit region r_i is $mix_i = \eta_i E_i^L + \zeta_i E_i^{ID}$, where η_i and ζ_i are location density and identity density, correspondingly.

In Fig. 5.6, we compare the privacy loss of the three algorithms when facing the same dynamic adversary with limited computing ability. The proposed and short-sighted algorithms have better performances than the static algorithm, while the performances become better and better with the increase of iteration times. Moreover, the proposed algorithm is superior to the shortsighted algorithm. This is because

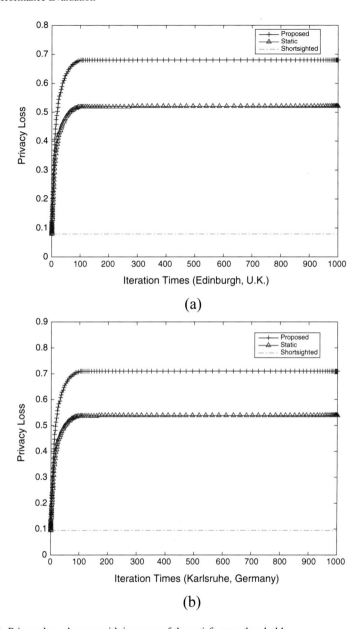

Fig. 5.6 Privacy loss changes with increase of the satisfactory threshold

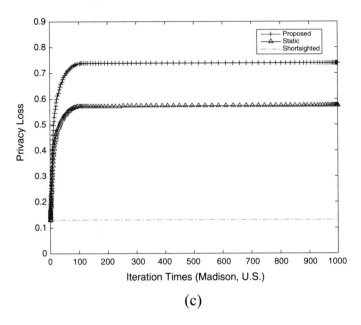

(c)

Fig. 5.6 (continued)

the static algorithm fixes its strategies while the shortsighted algorithm does not consider the overall privacy loss (with future privacy loss) as a whole. In addition, with the increase of the data size, we can see that the iteration times maintains the same level while the privacy loss decreases. The reason is that a larger density leads to less privacy leakage.

In Fig. 5.7, we can see that the privacy loss increases with the decrease of the percentage of the sensitive messages. The privacy loss of the static algorithm is relative high and fast converge to a higher constant, while the other two algorithms start from a relative low level and converge to a much lower constant than the static algorithm. The proposed algorithm has the best performances for the dynamic strategy and taking future payoff into account. The trends show that less percentage of messages leads to less privacy leakage. Analogy to the impact of iteration times, data size has a positive impact on the privacy loss for the same reason.

5.7.4 Convergence Speed

We present the comparisons of convergence speed between the modified SARSA algorithm and the classic algorithm among this subsection. Figure 5.8 shows the CDF of iterations for the user to obtain his best strategy in three different cities in different countries.

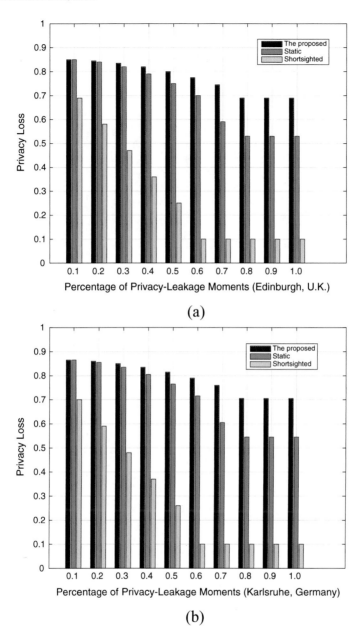

Fig. 5.7 Privacy loss changes with increase of the satisfactory threshold

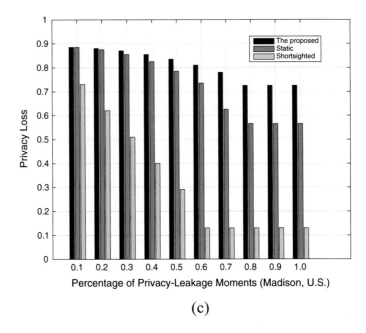

(c)

Fig. 5.7 (continued)

The iteration times of the proposed algorithm are less than 600, 1000, and 1100 respectively, while the iteration times of the classic algorithm are all between 10^5 and 10^6. These results clearly indicate that the proposed algorithm has a much faster convergence speed than the classic algorithm. The acceleration is due to the significant decrease of the cardinality, which is reduced to 2 in our algorithm.

5.8 Summary and Future Work

The wide spread of mobile devices has significantly promoted the development of CPSN. However, the information released by users contains sensitive information for improving utility, which puts the location privacy and identity privacy under great threats. In this chapter, we propose a method to provide hybrid privacy protection while maintaining high-quality data utilities. We model the actions series of the user and the adversary as a multi-stage zero-sum dynamic game while the long-term attack-defense process is formulated by a GMDP. We discuss about the payoff of the user and then formulate the problem as the user wants to maximize his overall payoff while the adversary holds the opposite target. Moreover, we propose a modified SARSA reinforcement learning algorithm to achieve fast convergence to the unique NE point, which is the best strategy for the user. This is an early work to provide hybrid privacy protection model in a long-term situation. At last, we use theoretical

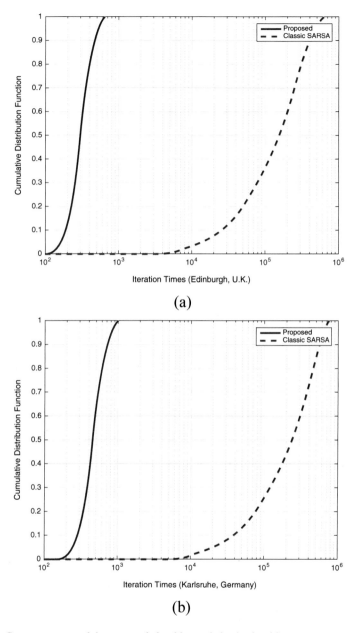

(a)

(b)

Fig. 5.8 Convergence speed the proposed algorithm and classic algorithm

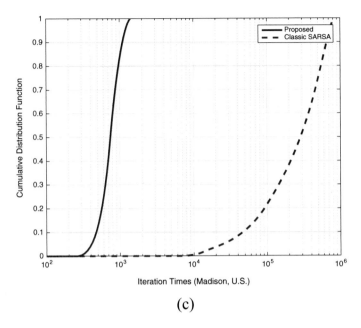

(c)

Fig. 5.8 (continued)

analysis and experiments on real-world datasets to prove the effectiveness of our proposed model.

For future work, we have the following ideas. On one hand, we will focus on the methods to provide more comprehensive privacy protection in various scenarios. On the other hand, we will keep trying to design a more uniform measurements for both privacy loss and data utility. Moreover, the customized privacy protection is another promising direction to research, which could be further improved by the combination of cross-discipline subjects.

References

1. Y. Qu, S. Yu, L. Gao, W. Zhou, S. Peng, A hybrid privacy protection scheme in cyber-physical social networks. IEEE Trans. Comput. Soc. Syst. **5**(3), 773–784 (2018)
2. B. S. Gu, L. Gao, X. Wang, Y. Qu, J. Jin, S. Yu, Privacy on the edge: Customizable privacy-preserving context sharing in hierarchical edge computing. IEEE Trans. Netw. Sci. Eng. (2019)
3. B. Gu, X. Wang, Y. Qu, J. Jin, Y. Xiang, L. Gao, Context-aware privacy preservation in a hierarchical fog computing system, in *ICC 2019-2019 IEEE International Conference on Communications (ICC)* (IEEE, 2019), pp. 1–6
4. X. Wang, B. Gu, Y. Qu, Y. Ren, Y. Xiang, L. Gao, A privacy preserving aggregation scheme for fog-based recommender system, in *International Conference on Network and System Security* (Springer, 2020), pp. 408–418

5. X. Wang, B. Gu, Y. Qu, Y. Ren, Y. Xiang, L. Gao, Reliable customized privacy-preserving in fog computing, in *ICC 2020-2020 IEEE International Conference on Communications (ICC)* (IEEE, 2020), pp. 1–6
6. A. Sheth, P. Anantharam, C. Henson, Physical-cyber-social computing: an early 21st century approach. IEEE Intell. Syst. **28**(1), 78–82 (2013)
7. S. Yu, M. Liu, W. Dou, X. Liu, S. Zhou, Networking for big data: a survey. IEEE Commun. Surv. Tutor. **19**(1), 531–549 (2017)
8. W. Wang, Q. Zhang, Privacy preservation for context sensing on smartphone. IEEE/ACM Trans. Netw. **24**(6), 3235–3247 (2016)
9. Q. Xu, P. Ren, H. Song, Q. Du, Security-aware waveforms for enhancing wireless communications privacy in cyber-physical systems via multipath receptions. IEEE Internet Things J. **PP**(99) (2017)
10. A.R. Beresford, F. Stajano, Location privacy in pervasive computing. IEEE Pervasive Compu. **2**(1), 46–55 (2003)
11. X. Zheng, Z. Cai, J. Yu, C. Wang, Y. Li, Follow but no track: privacy preserved profile publishing in cyber-physical social systems. IEEE Internet Things J. (2017)
12. G. Tong, W. Wu, S. Tang, D. Du, Adaptive influence maximization in dynamic social networks. IEEE/ACM Trans. Netw. **25**(1), 112–125 (2017)
13. L. Tang, X. Chen, S. He, When social network meets mobile cloud: a social group utility approach for optimizing computation offloading in cloudlet. IEEE Access **4**, 5868–5879 (2016)
14. S. Yu, Big privacy: challenges and opportunities of privacy study in the age of big data. IEEE Access **4**, 2751–2763 (2016)
15. T. Maekawa, N. Yamashita, Y. Sakurai, How well can a user's location privacy preferences be determined without using GPS location data? IEEE Trans. Emerg. Topics Comput. **PP**(99), 1 (2016)
16. Y. Zheng, W. Feng, P. Wang, Anonymous authentication for trustworthy pervasive social networking. IEEE Trans. Comput. Soc. Syst. **2**(3), 88–98 (2015). https://doi.org/10.1109/TCSS. 2016.2519463
17. E. Shen, T. Yu, Mining frequent graph patterns with differential privacy, in *Proceedings of KDD 2013, Chicago, IL, USA, August* 11–14, 2013 (2013), pp. 545–553
18. H. Liu, X. Li, H. Li, J. Ma, X. Ma, Spatiotemporal correlation-aware dummy-based privacy protection scheme for location-based services, in *Proceedings of the IEEE INFOCOM 2017, Atlanta, GA, USA, May* 1–4, 2017 (2017)
19. K. Vu, R. Zheng, J. Gao, Efficient algorithms for k-anonymous location privacy in participatory sensing, in *Proceedings of the IEEE INFOCOM 2012, Orlando, FL, USA, March* 25–30, 2012 (2012), pp. 2399–2407
20. R. Shokri, G. Theodorakopoulos, C. Troncoso, J. Hubaux, J. Boudec, Protecting location privacy: optimal strategy against localization attacks, in *The ACM Conference on Computer and Communications Security, CCS'12, Raleigh, NC, USA, October* 16–18, 2012 (2012), pp. 617–627
21. Y. Mao, H. Shen, Web of credit: adaptive personalized trust network inference from online rating data. IEEE Trans. Comput. Soc. Syst. **3**(4), 176–189 (2016)
22. S. Pierangela, S. Latanya, Protecting privacy when disclosing information: k-anonymity and its enforcement through generalization and suppression, in *Proceedings of the IEEE Symposium on Research in Security and Privacy* (1998), pp. 1–19
23. A. Machanavajjhala, D. Kifer, J. Gehrke, M. Venkitasubramaniam, *L*-diversity: Privacy beyond *k*-anonymity. IEEE Trans. Knowl. Data Eng. **1**(1) (2007)
24. N. Li, T. Li, S. Venkatasubramanian, Closeness: a new privacy measure for data publishing. IEEE Trans. Knowl. Data Eng. **22**(7), 943–956 (2010)
25. Y. Qu, S. Yu, L. Gao, J. Niu, Big data set privacy preserving through sensitive attribute-based grouping, in *IEEE International Conference on Communications, ICC 2017, Paris, France, May* 21–25, 2017 (2017), pp. 1–6
26. Y. Qu, S. Yu, L. Gao, S. Peng, Y. Xiang, L. Xiao, Fuzzydp: fuzzy-based big data publishing against inquiry attacks, in *2017 IEEE Conference on Computer Communications Workshops, INFOCOM Workshops, Atlanta, GA, USA, May* 1–4, 2017 (2017), pp. 7–12

27. Y. Qu, L. Cui, S. Yu, W. Zhou, J. Wu, Improving data utility through game theory in personalized differential privacy, in *IEEE International Conference on Communications, ICC 2018, Kansas City, America, May* 21–25, 2018 (2018), pp. 1–6

28. C. Dwork, Differential privacy, in *in Proceedings of ICALP 2006, Venice, Italy, July 10-14, 2006, Proceedings, Part II* (2006), pp. 1–12

29. M. Groat, W. He, S. Forrest, KIPDA: k-indistinguishable privacy-preserving data aggregation in wireless sensor networks, in *in Proceedings of the IEEE INFOCOM 2011, Shanghai, China, April* 10–15 2011 (2011), pp. 2024–2032

30. Y. Gong, C. Zhang, Y. Fang, J. Sun, Protecting location privacy for task allocation in ad hoc mobile cloud computing. IEEE Trans. Emerg. Topics Comput. **PP**(99), 1 (2016)

31. B. Liu, W. Zhou, T. Zhu, L. Gao, T. Luan, H. Zhou, Silence is golden: enhancing privacy of location-based services by content broadcasting and active caching in wireless vehicular networks. IEEE Trans. Vehicular Technol. **65**(12), 9942–9953 (2016)

32. L. Zhang, C. Hu, Q. Wu, J. Domingo-Ferrer, B. Qin, Privacy-preserving vehicular communication authentication with hierarchical aggregation and fast response. IEEE Trans. Comput. **65**(8), 2562–2574 (2016)

33. M. Nasim, R. Charbey, C. Prieur, U. Brandes, Investigating link inference in partially observable networks: friendship ties and interaction. IEEE Trans. Comput. Soc. Syst. **3**(3), 113–119 (2016)

34. F. Koufogiannis, G.J. Pappas, Diffusing private data over networks. IEEE Trans. Control Netw. **PP**(99), 1 (2017)

35. J. Freudiger, M. Manshaei, J. Hubaux, D. Parkes, On non-cooperative location privacy: a game-theoretic analysis, in *Proceedings of the 2009 ACM Conference on Computer and Communications Security, CCS 2009, Chicago, Illinois, USA, November 9–13, 2009* (2009), pp. 324–337

36. K. Zhang, X. Liang, R. Lu, X. Shen, PIF: a personalized fine-grained spam filtering scheme with privacy preservation in mobile social networks. IEEE Trans. Comput. Soc. Syst. **2**(3), 41–52 (2015)

37. Y. Qu, M. R. Nosouhi, L. Cui, S. Yu, Privacy preservation in smart cities, in *Smart Cities Cybersecurity and Privacy* (Elsevier, 2019), pp. 75–88

38. Y. Qu, S. Yu, W. Zhou, S. Peng, G. Wang, K. Xiao, Privacy of things: emerging challenges and opportunities in wireless internet of things. IEEE Wirel. Commun. **25**(6), 91–97 (2018)

39. J. Yu, K. Wang, D. Zeng, C. Zhu, S. Guo, Privacy-preserving data aggregation computing in cyber-physical social systems. ACM Trans. Cyber-Phys. Syst. **3**(1), 1–23 (2018)

40. L. Cui, G. Xie, Y. Qu, L. Gao, Y. Yang, Security and privacy in smart cities: challenges and opportunities. IEEE Access **6**, 46 134–46 145 (2018)

41. L. Cui, Y. Qu, L. Gao, G. Xie, S. Yu, Detecting false data attacks using machine learning techniques in smart grid: a survey. J. Netw. Comput. Appl. 102808 (2020)

42. L. Gao, T.H. Luan, B. Gu, Y. Qu, Y. Xiang, *Privacy-Preserving in Edge Computing Wireless Networks, Ser* (Springer, 2021)

43. L. Gao, T. H. Luan, B. Gu, Y. Qu, Y. Xiang, Blockchain based decentralized privacy preserving in edge computing, in *Privacy-Preserving in Edge Computing* (Springer, 2021), pp. 83–109

44. L. Gao, T. H. Luan, B. Gu, Y. Qu, Y. Xiang, Context-aware privacy preserving in edge computing, in *Privacy-Preserving in Edge Computing* (Springer, 2021), pp. 35–63

45. L. Gao, T. H. Luan, B. Gu, Y. Qu, Y. Xiang, An introduction to edge computing, in *Privacy-Preserving in Edge Computing* (Springer, 2021), pp. 1–14

46. L. Gao, T. H. Luan, B. Gu, Y. Qu, Y. Xiang, Privacy issues in edge computing, in *Privacy-Preserving in Edge Computing* (Springer, 2021), pp. 15–34

47. Y. Qu, L. Gao, Y. Xiang, Blockchain-driven privacy-preserving machine learning, in *Blockchains for Network Security: Principles, Technologies and Applications* (2020), pp. 189–200

48. Y. Qu, M. R. Nosouhi, L. Cui, S. Yu, Personalized privacy protection in big data

49. Y. Qu, M. R. Nosouhi, L. Cui, S. Yu, Existing privacy protection solutions, in *Personalized Privacy Protection in Big Data* (Springer, 2021), pp. 5–13

50. Y. Qu, M. R. Nosouhi, L. Cui, S. Yu, Future research directions, in *Personalized Privacy Protection in Big Data* (Springer, 2021), pp. 131–136
51. Y. Qu, M. R. Nosouhi, L. Cui, S. Yu, Leading attacks in privacy protection domain, in *Personalized Privacy Protection in Big Data* (Springer, 2021), pp. 15–21
52. Y. Qu, M. R. Nosouhi, L. Cui, S. Yu, Personalized privacy protection solutions, in *Personalized Privacy Protection in Big Data* (Springer, 2021), pp. 23–130
53. M. Alsheikh, D. Hoang, D. Niyato, H. Tan, S. Lin, Markov decision processes with applications in wireless sensor networks: a survey. IEEE Commun. Surv. Tutor. **17**(3), 1239–1267 (2015)
54. L. Kaelbling, M. Littman, A. Moore, Reinforcement learning: a survey. J. Artif. Intell. Res. **4**, 237–285 (1996)
55. B. Wang, Y. Wu, K. Liu, T. Clancy, An anti-jamming stochastic game for cognitive radio networks. IEEE J. Selected Areas Commun. **29**(4), 877–889 (2011)
56. M. Götz, S. Nath, J. Gehrke, Maskit: privately releasing user context streams for personalized mobile applications, in *Proceedings of the ACM SIGMOD International Conference on Management of Data, SIGMOD 2012, Scottsdale, AZ, USA, May 20-24, 2012* (2012), pp. 289–300
57. Y. Center, Yelp dataset 8th round, https://www.yelp.com/dataset_challenge

Chapter 6
Future Research Directions

From Chaps. 2–5, we have shown the existing research status of machine learning driven privacy preservation in IoTs, especially focus on three leading directions using GAN, federated learning, and reinforcement learning. Several advanced technologies and theories are also integrated, for example, differential privacy, game theory, blockchain, etc. Nevertheless, there are still plenty of significant and prospective issues worthy investigating. The popularization of blockchain, digital twin, and artificial intelligence offers a mass of opportunities for researches on machine learning driven privacy preservation in IoTs, but meanwhile they raise new challenges such as unsatisfying data utility and limited communication and computing resources. In addition, there are various other research topics that desiderata consideration in machine learning driven privacy preservation in IoTs. To pave the way for readers and forthcoming researchers, we outline several potentially promising research directions that may be worthy of future efforts.

6.1 Trade-Off Optimization in IoTs

Trade-off between privacy protection and data utility has always been one of the primary targets in privacy preservation field. The reason is as explained above. In privacy preservation domain, data curators will publish sanitized data to all data requestors by conducting irreversible data distortion in most cases. In this case, data utility should be well-considered for the value of the sanitized data.

However, most existing research barely considers limited computing and communication resources, which should be taken into account in IoT scenarios. That makes the trade-off optimization even more complicated. Currently, there are two potential solutions, including both a static model and a dynamic model.

For the static solution, it requires that the designer is experienced in this domain and thereby can identify the issues in each procedure. By deeply understanding such a system, an expert may be able to design a novel loss function that can help improve the performance of the machine learning based privacy preservation models and achieve fast convergence under the constrain of limited computing and communication resources. Although the design of a loss function can improve the overall performance, but it is usually case-by-case and can hardly be generalized.

A dynamic solution is usually in a form of a iterated algorithm, for instance, a machine learning algorithm or a game theory based model. The optimization process can be formulated as a Markov Decision Process. To achieve this, all parties' actions, states, and corresponding payoff functions should be modelled. In this case, the payoff function can be modelled as the trade-off between privacy protection, data utility, and resource limitations. The Markov Decision Process can be solved using various ways, such as Q-learning algorithm or state-action-reward-state-action (SARSA) algorithm. However, the adoption of dynamic solutions may bring about extra burden on computing and communication resources. Therefore, the efficiency improvement is also worth further investigating.

6.2   Privacy Preservation in Digital Twined IoTs

Digital twin is a fast emerging technology in recent years. Originated from manufacture industries, digital twin has proved its effectiveness when designing and improving specific physical objects and their interactions. This create opportunities to IoTs as well. In an IoT network, there are various IoT devices such as sensors, cameras, edge devices, vehicles, etc. It is possible to create digital twins for IoT devices in a cloud server to simulate an IoT environment, which has been evidenced in a mass of existing research.

The establishment of digital twined IoT enables the optimization of execution policies of IoT devices, communication efficiency, and a lot more. However, privacy issues emerge since local data of IoT devices should be uploaded to the digital twin. Besides, the digital twins are always located in a server with high computing, communication, and storage devices, such as a cloud server. This poses further challenges to privacy protection due to the existence of multiple attacks like man-in-the-middle attacks.

To preserve the privacy in this case, it is possible to perform federated meta learning. Meta-learning, or learning to learn, is the science of systematically observing how different machine learning approaches perform on a wide range of learning tasks, and then learning from this experience, or meta-data, to learn new tasks much faster than otherwise possible. As is known to all, federated learning is able to achieve privacy-preserving distributed machine learning by exchanging the model parameters or gradients of a commonly-maintained model. By storing local data locally, privacy protection is achieved. However, federated learning meets several bottlenecks, of which a primary challenge is to improve the model performance. Therefore, meta learning can be accommodated in federated learning such that the performance and privacy protection can be balanced. Similar the other IoT scenarios, the adoption of machine learning models result in the consumption of computing and communication resources, which should be considered while establishing such a model.

6.3   Personalized Consensus and Incentive Mechanisms for Blockchain-Enabled Federated Learning in IoTs

6.4   Privacy-Preserving Federated Learning in IoTs

Consensus algorithm for B-FL: The consensus algorithm is the core of any blockchain system, and an ideal consensus algorithm should be computationally efficient and highly secure. PoW and PoS are the two most widely used consensus algorithms but the former is extremely inefficient and costly due to the use of a nonce-finding mechanism, while the later will weaken the decentralization property of blockchain as the miners with a huge number of stakes can dominate the blockchain system. To overcome these drawbacks, we propose to directly use the B-FL mission as a consensus proof, without running a separate consensus proof process nor utilising the miners' stakes as a consensus proof.

Based on the proposed B-FL process above, we will develop the new consensus algorithm in the following way. In Phase 1, each participating miner uses their local data to train a local model. Once completing the local model training, they will broadcast the local model parameters to other participating miners. When a predefined percentage of miners have completed their local model training or a predefined time length is reached, Phase 1 ends. All the miners who have successfully completed their work are eligible for participating in Phase 2 and those miners who did not complete Phase 1 will be excluded from being involved in the next phase. In Phase 2, each miner eligible for this phase will execute a smart contract containing a model verification algorithm and a model selection algorithm to verify the authenticity of all local models and select those authentic local models suitable for aggregation. Phase 2 will finish once a predefined percentage of miners have completed their work or a predefined time length is reached. The miners whose local models have been selected for aggregation will proceed to Phase 3 and those whose local models are classified as falsified or unsuitable for aggregation will be excluded from participating in the next phase. In Phase 3, each miner eligible for this phase will execute a smart contract containing a model aggregation algorithm to aggregate the selected local models to generate their global model. Then the miner will store the generated global model parameters into a candidate block and broadcast it to the blockchain system. When a predefined percentage of miners have completed their work or a predefined time length is reached, each miner eligible for this phase will vote for a (different) global model whose model parameters are the closest to their own. If two or more models have the same closeness to their own, the one that was generated the earliest will be chosen. After a predefined time period, the global model receiving the highest number of votes is elected as the final global model in that B-FL round. The associated miner is recognised as the winning miner of that B-FL round and the associated candidate block is formally appended on the blockchain system, which ends the consensus process. As an inherent part of the consensus algorithm development, we will analyse the robustness of the proposed consensus algorithm against attacks and study how the algorithmic parameters, such as those predefined percentages of miners and predefined time lengths, will affect its performance. Based on the findings, we will refine the proposed consensus algorithm accordingly.

Incentive algorithm for B-FL: The ultimate objective of B-FL is to motivate miners to actively participate in the B-FL process to collectively produce high-quality models within the shortest possible timeframe. The incentive mechanisms used in traditional blockchain systems are not suitable for B-FL as they only award the winning miner. Intuitively, it is better to devise a personalized incentive algorithm to award all the miners who have made valuable contributions to B-FL. The proposed awarding principles are as follows. Firstly, in each B-FL round, every miner whose local model was selected for aggregation will be awarded and the award amount depends on the quality of the model (the higher quality the better) and the time spending on training the model (the shorter time the better). This awarding principle is reasonable as producing a high-quality local model in a short timeframe requires the miner to use more high-quality data and computing power to train their local model in Phase 1 and use more computing power to perform local model verification and selection in Phase 2. Secondly, the miners who participated in the whole Phase 3 to determine the final global model and the winning miner will receive an additional award. Thirdly, the winning miner will receive a further top-up award. In addition to these awarding principles, certain constraints should be imposed to avoid the over-fitting issue when determining the award amount since over-fitting will weaken the extensiveness to diverse data from different miners. Based on the proposed awarding principles and constraints, an award mapping algorithm will be proposed to map the miners' contributions to the actual rewards to be allocated. We will also investigate how to optimize the award mapping algorithm by considering its impact on model accuracy, model training convergence speed, miner participation rate, etc.

6.5   Federated Generative Adversarial Network in IoTs

In this section, we discuss our plan on federated generative adversarial network for IoT applications in practice.

The basic idea is that a central server, usually a cloud server or a edge server, servers as the Discriminator. At the same time, each IoT end device works as a Generator and generates synthetic data using its local data. The synthetic data is then uploaded to the central server for discrimination. As the central server has a full vision of all generated synthetic data, it can perform discrimination in a powerful way. Then all the local Generators game with a central Discriminator iteratively until convergence.

The difficult part of this model includes the high communication overhead since the synthetic data should be uploaded to the central server. Besides, the synthetic data may leak privacy information during transmission.

Therefore, another potential solutions is that each IoT device trains a GAN model locally and send the GAN model parameters to a central server for aggregation. This can significantly reduce the communication overhead but requires certain degree of computation resources of IoT devices.

Therefore, there should be a trade-off between these two methods. That is also part of our research plan in the future.

Chapter 7
Summary and Outlook

In this monograph, we summarize the state-of-art research on machine learning driven privacy protection models in IoT scenarios. As far as we can see, machine learning driven privacy protection is still in its early stage. What have been present in this monograph, like models, theories, and potentially conceptual designs, could serve as a point of departure for follow-up readers, students, engineers, and researchers to investigate this emerging field. Our target is to provide a systematic summary of existing research and application outputs on machine learning driven privacy protection in IoTs. Based on this, we analyze the theoretical and practical applicability under big data settings. Subsequently, we offer the interested readers several future research directions, which we hope the following explorers find them insightful or inspiring from a certain angle.

In short, plenty of existing research has been surveyed and compared to offer a big picture of the research and application prospect of machine learning driven privacy protection. We identify several key issues that prevent it from further development, which includes trade-off, unintended privacy leakage, etc. To better demonstrate, we discuss the machine learning driven privacy protection from perspectives federated learning, generative adversarial network, and reinforcement learning. The illustrated models can be generalized to IoTs or a wider range of application scenarios.

Beyond that, this monograph contains multiple popular theories to improve machine learning driven privacy protection. They are game theory, blockchain, differential privacy, cryptography methods, etc. We correspondingly provide analysis on their feasibility while adopting into various real-world IoT practices.

Built upon the existing research and obtained results, we further discuss several future research directions, in particular, trade-off optimization in IoTs, privacy preservation of digital twins in IoTs, privacy-preserving blockchain-enabled federated learning in IoTs, privacy-preserving federated learning in IoTs, federated generative adversarial network in IoTs.

Y. Qu et al., *Privacy Preservation in IoT: Machine Learning Approaches*, SpringerBriefs in Computer Science, https://doi.org/10.1007/978-981-19-1797-4_7

As we discussed above, machine learning driven privacy protection is a emerging research domain that is under-explored. So far, researchers and public raises more questions than answers. In the future, there will be a crowd of other unprecedented problems, and many unknown of unknowns. Built up on our existing research, we intend to share several focal points with passionate researchers and readers as follows.

In the first place, machine learning driven privacy preservation in IoTs requires the integration of different disciplines, including but not limited to information theory, legislation, social science, etc. We are quite sure that computer science is not able to deal with all privacy issues by itself. In accordance with the scientific methodology, the measurement of privacy comes first, which is followed by mathematical modeling, idea confirmation, as well as conclusion drawing from both experimental and theoretical results and analysis. However, there is not existing research to measure the privacy in a generalized and applicable way. It is intuitive that privacy is difficult to evaluate like several feeling concepts including sadness and anger. As Sir Issac Newton once complained, he already can calculate the movement of stars but still fail to measure the madness of human-beings. Even after several centuries, the similar difficulty is still there. Fortunately, the society is experiencing a booming of all disciplines under the framework of science. It is believed that the horizon is just ahead for us, and we need to seize the opportunity to integrate diverse skills to make it happen.

It is indeed that cross-discipline research is attractive, but requires unimaginable efforts to perform. Based on a statistic result of Science Magazine, almost over 60% of articles published by Science Magazine are cross-discipline research. The result can show a strong link between powerful research and cross-discipline skills. Despite this, the difficulties of performing it is obvious according to our experience. Usually, within the time of a cup of coffee, researchers from different disciplines may produce several inspiring ideas, but it is particularly difficult for the team to move on since people talk in different "languages". Therefore, a general advice is that researchers should master the skills from other disciplines instead of posing questions to collaborators for answers.

In the second place, we need theoretical "weapons" for privacy research. Till now, in the machine learning domain, GAN and federated learning are the only techniques that can guarantee privacy protection during machine learning model training. However, GAN is not dedicated for privacy protection while federated learning can only provide privacy protection when machine learning is required and data is distributed. These scenarios are a tiny proportion of privacy protection in IoT scenarios. There are several potential opportunities listed in the previous chapters, but can be summarized in two directions.

- Upgrade existing tools: Upgrade existing tools is the most efficient way to enhance privacy protection in a specific domain. Standing on the shoulders of giants, researchers are able to see further and target higher.
- Invent new tools: Inventing new tools is a tough but promising job. We can see new tools emerging every several years and then overtaken or enhanced by newer tools.

This brief monograph is the concise summary of our research outputs in recent 5 years, which is also an initial step of our research group. We sincerely wish the shallow effort we made can draw interested readers' attention to investigate this promising land with us, no matter you are an engineer, a student, a research amateur, or anyone.

In the end, I and my team members really appreciate your patience and interest in this book. We look forward to your comments, feedback, or suggestions. Moreover, we sincerely anticipate reading your future works in this domain.

Printed in the United States
by Baker & Taylor Publisher Services